Emergence
and Embodiment

SCIENCE AND CULTURAL THEORY

A Series Edited by Barbara Herrnstein Smith

and E. Roy Weintraub

Emergence and Embodiment

New Essays on Second-Order Systems Theory

EDITED BY BRUCE CLARKE AND MARK B. N. HANSEN

Duke University Press · *Durham & London* · 2009

© 2009 Duke University Press
All rights reserved
Printed in the United States
of America on acid-free paper ∞
Designed by Amy Ruth Buchanan
Typeset in Minion by
Achorn International Inc.
Library of Congress Cataloging-in-
Publication Data appear on the
last printed page of this book.

Contents

Acknowledgments

The groundwork for this volume was a set of conference panels Bruce Clarke organized for the third international meeting of the Society for Literature, Science, and the Arts, held June 2004 in Paris, France. The panelists for "Neocybernetic Emergence I and II" were Bruce Clarke, Steven Meyer, Mark Hansen, Eric White, Edgar Landgraf, and Cary Wolfe.

We would like to thank all the participants in this volume for their contributions. We are happy to acknowledge Thomas von Foerster for his kind approval of the text of the interview with Heinz von Foerster and Amanda Pask Heitler for her permission to republish images created by Gordon Pask for von Foerster's "On Self-Organizing Systems and Their Environments." We are grateful to Carl-Auer-Systeme Verlag for permission to publish a translated excerpt of *Einführung in die Systemtheorie*, by Niklas Luhmann, and to Amy Cohen-Varela and John Wiley and Sons, Ltd., for permission to republish Francisco J. Varela's memoir, "The Early Days of Autopoiesis," originally pub-lished in *Systems Research*.

It has been a pleasure to work with Reynolds Smith and the staff at Duke University Press. Finally, we would like to acknowledge the assistance and support of Yves Abrioux, Barbara Herrnstein-Smith, Timothy Lenoir, Lynn Margulis, Robert Markley, Albert Müller, Manuela Rossini, and William Irwin Thompson.

Introduction

Neocybernetic Emergence

BRUCE CLARKE AND MARK B. N. HANSEN

In his introduction to *Observing Systems*, "The Ages of Heinz von Foerster," first published in 1981, Francisco Varela concluded with a characterization of "the last age of Heinz."[1] In the chronology of second-order cybernetics, this would be considered its *first* age, the period during the early 1970s when von Foerster laid out his ground-breaking sketches of, in Varela's words, "recursive mechanisms in cognitive systems," thereby producing the initial formulations for a cybernetics of cybernetics.[2] What struck Varela in the early 1980s was the extent to which the force of von Foerster's cognitive innovations had not yet gained secure footholds in the mainstream academy, had "*not* permeated our intellectual preferences and current thinking":

> There is little doubt that our current models about cognition, the nervous system, and artificial intelligence are severely dominated by the notion that information is represented from an out-there into an in-here, processed, and an output produced. There is still virtually no challenge to the view of objectivity understood as the condition of independence of descriptions, rather than a circle of mutual elucidation. Further, there is little acceptance yet that the key idea to make these points of view scientific programmes is the operational closure of cognizing systems, living or otherwise. These are precisely the leitmotives of Heinz's last stage.[3]

Since Varela made this observation, there has certainly been some significant, if modest, penetration of these fundamental cognitive motifs into the "intellectual preferences" of thinkers across the spectrum of natural, mathematical, and discursive disciplines. As we see it, however, Varela's words still ring true of our present time, and to the extent that they do, this volume of essays has important work to do. For it is only by theorizing the operational closure of cognizing systems that cultural theory can rescue agency—albeit agency of a far more complex variety than that of traditional humanism—from being overrun by the technoscientific processes that are everywhere transforming the material world in which we live today. Indeed, given the acceleration in technoscientific development since the 1980s—acceleration that has witnessed the advent of artificial life, complexity theory, and other technosciences of emergence—the imperative to theorize the operational closure of cognizing systems has, arguably, never been more urgent. Better late than never, second-order cybernetics can perhaps now finally come through on its promise to provide the ecology of mind best fitted to the demands of our intellectual, institutional, and global crises.

From Cybernetics to Neocybernetics

The cultural history of cybernetics is still being written. There is no authoritative version but rather a swarm of competing accounts. Given the welter of disciplines engaged in the movement, as well as the self-reflexive turn in cybernetic thought itself, a definitive history would be an impossible project. As has often been told, however, the first cybernetics emerged in the 1940s as a technoscience of communication and control, drawing from mathematical physics, neurophysiology, information technology, and symbolic logic. Historically concurrent with the postwar spread of linguistic structuralism in Europe, cybernetics was set forward in the United States and then vigorously transplanted to Soviet and European subcultures. From a base connecting biological and computational systems by way of information theory and communications technology, cybernetics was academically mainstreamed under the names Artificial Intelligence (AI) and, more broadly, computer science in the service of command-and-control systems. But due to the long interdisciplinary roster of Warren McCulloch's invitees to the Macy Conferences—including Lawrence Frank, Heinrich Klüver, Gregory Bateson, Margaret Mead, and Lawrence Kubie—cybernetic discourse entered psychology, anthropology, and other social sciences and from there, in the 1950s and '60s, the humanities and the creative arts.

Coined by original Macy participant Norbert Wiener, "cybernetics" in its initial formulation was the "study of messages, and in particular of the effective messages of control."[4] But for Wiener cybernetics also raised new issues about the "definition of man": If "human behaviors" can be duplicated by machines, how is one to "differentiate man" from other entities?[5] Keeping the focus on information and communication but extending it beyond machines, Wiener argued that among living beings only "man" is obsessively driven to communicate. While this was not in fact a satisfactory criterion of distinction between human beings and other living and nonliving things, it did show that from the start, cybernetics put the ontology of "humanity" into question.

Less than a decade later, W. Ross Ashby deflected Wiener's emphasis on human communication and control toward the ontological neutrality of Claude Shannon's information theory.[6] In a similar vein, Gregory Bateson wrote of the first cybernetics that its "subject matter" extended across traditional disciplinary registers in focusing on "the propositional or informational aspect of the events and objects in the natural world."[7] Now we would say that cybernetic methodologies draw out the virtuality correlated with actuality, but clearly the shift in emphasis from the actual to the virtual was already under way in first-order cybernetics. According to Ashby, with regard to the substance of the media conveying informatic forms, "the materiality is irrelevant."[8] Cybernetics marks a shift away from the building blocks of phenomena—so long the focus of chemistry and physics and, given the success of these disciplines, too often a model for biology and psychology—to the form of behaviors, what things do and how they are observed.

The first-order cybernetic demotion of material substance relative to informatic pattern is memorably recorded in *The Human Use of Human Beings*, when Wiener rehearsed a teleporter scenario that characteristically operated on a biased form/substance binary: "The individuality of the body is that of a flame rather than that of a stone, is that of a form rather than that of a bit of substance."[9] Nevertheless, the teleportation of that form for the purpose of rematerialization at a distance would almost surely involve at least the momentary destruction of the organic being undergoing the process. One service of the science fictions retailed as *The Fly* has been to restore the visceral horror of a process that Wiener described with a remarkably bloodless and surgical élan:

> Any scanning of the human organism must be a probe going through all parts, and must have a greater or less tendency to destroy the tissue on its way. To hold an organism stable while part of it is being slowly destroyed, with the intention of recreating it out of other material elsewhere, involves a lowering of its degree of activity, which in most cases we should consider to prevent life in the tissue. In other words, the fact that we cannot telegraph the pattern of a man from one place to another is probably due to technical difficulties, and in particular, to the difficulty of keeping an organism in

being during such a radical reconstruction. It is not due to any impossibility of the idea.[10]

One could generalize from Wiener and Ashby—as well as from much of its popular offspring in cyberpunk and other technoid fantasies—that first-order cybernetics remains inscribed within classical scientific thought: it holds onto humanist and idealist dualisms that describe the world in terms of an equivocal dialectics of matter and form, of substance and pattern, in which the immaterial wrests agency away from the embodied.

One way to mark the emergence of neocybernetics (our preferred paraphrase of "second-order cybernetics" or "the cybernetics of cybernetics") is to emphasize its new questioning and eventual overcoming of classical substance/form distinctions. Neocybernetic systems theory radicalizes the constructivist epistemology inscribed within the first cybernetics by shifting to an autological rather than ontological theory of form. In neocybernetic theory, the form/substance dichotomy is superseded by the distinction between form and *medium*.

Putting form and medium theory together, neocybernetics goes beyond classical ontology's impasse—is it form, or is it matter, nothing, or everything?—before the oscillations of being and nonbeing. Such imponderables have always presupposed some ultimate fundament upon which to evaluate this all-or-nothing conundrum. Neocybernetic epistemology replies by "de-ontologizing" the question. Neither form nor medium reaches bottom: there is no bottom. Forms are temporary fixations of elements within a medium, and when enough like forms coalesce, they become another medium for a new, emergent set of forms. As Edgar Landgraf discusses below in the context of improvisational performances, this autological dynamic is especially decisive in the social communication of art forms. In "The Medium of Art," Niklas Luhmann writes: "In the case of art . . . form first constitutes the medium in which it expresses itself. Form is then a 'higher medium,' a second-degree medium which is able to use the difference between medium and form itself in a medial fashion as a medium of communication."[11] Building upon this understanding, Michael Schiltz in this volume draws out the implications for Luhmann's theory of positing *meaning* as the medium within which psychic and social forms interpenetrate: "If the medium of meaning is indeed the ultimate medium of psychic and social systems—i.e., if meaning is 'the medium of itself'—then what is its 'form,' the distinction through which it can be expressed? I perceive only one answer: the medium of meaning must be identical to the difference between form and medium, and the reentry of that distinction into itself. Its consequent indecidability is the symbol of our dealing with the world."

First-order cybernetics underscored the provisional nature or the *construct-edness* of cognitions within observing systems, but it did so by undercutting the significance and contribution of material/energetic environments to the cognitive systems that emerge within them. The strong constructivism of neo-cybernetic systems theory deals with the world by promoting a new level of attention to the media of its forms or, more concretely, to the *environments* and the *embodiments* of systems. As Bruce Clarke's essay demonstrates, we see this ecological convergence of constructivism and cybernetic "environmentalism" in the key figure responsible for the turn from first- to second-order systems theory, Heinz von Foerster. At the beginning of his 1974 essay "On Constructing a Reality," von Foerster, then director of the Biological Computer Laboratory at the University of Illinois, recounted how, "perhaps ten or fifteen years ago, some of my American friends came running to me with the delight and amaze-ment of having just made a great discovery: 'I am living in an Environment! I have always lived in an Environment! I have lived in an Environment through-out my whole life!' "[12] Yet despite the ecological revelation of their newfound Environment, according to von Foerster, his friends had yet to make another and even more crucial discovery: "When we perceive our environment, it is we who invent it."[13]

This collection contributes to the cultural work of cybernetic discourse by tracing the lines of neocybernetic development that extend directly from the work of Heinz von Foerster. Putting von Foerster at the head of neocybernetics throws attention on what is still a minority account of cybernetics' intellectual accomplishment and cultural significance. Neocybernetics' greatest interest for textual disciplines, media studies, and the social sciences, we argue, derives from particular advances upon first-order cybernetics in the biological, cognitive, and social systems theories developed in the work of von Foerster and Gregory Bate-son and extended from there by Henri Atlan, Humberto Maturana, Francisco Varela, Lynn Margulis, Susan Oyama, and Niklas Luhmann.[14]

Some of the most important theoretical and critical conversations going on today in the cognitive sciences, chaos and complexity studies, and social systems theory stem from neocybernetic notions of self-organization, emer-gence, and autopoiesis. A growing body of scholarly work is rethinking the shape and evolution of the relations among science, technology, sociology, psychology, philosophy, history, literature, and the arts through neocyber-netic terms. Expanding the initial transdisciplinary framework connecting the natural and human sciences with information technologies, recent thinkers, such as Michel Serres, Gilles Deleuze, Félix Guattari, Donna Haraway, Bruno Latour, and Isabelle Stengers, have deployed neocybernetic discourse extensively and transformatively. Neocybernetic discourse is central to current historical, interpretive, and theoretical investigations using concepts such as narrative, medium, assemblage, information, noise, network, and communication to remap the terrain of knowledge with reference to the operational boundaries of systems and their environments.

This body of work is both inspired and admonished by the larger unfolding of cybernetics, its institutional ups and downs, its cultural impacts and resistances, its culs-de-sac, and its continuing intellectual and social promise. Neocybernetic concepts in the line from von Foerster to Maturana, Varela, and Luhmann challenge not just the technoid rigidities of AI and first-order mechanical and social systems engineering, but also, and more profoundly, the epistemological foundations of philosophical humanism as such. Whether technical or biotic, psychic or social, systems are bounded semi-autonomous entities coupled *with* their environments and *to* other systems. One shifts attention from isolated elements and relations to the emergent behaviors of ever-larger ensembles. Neocybernetic systems theory stresses the recursive complexities of observation, mediation, and communication. Whatever comes to be (observed) owes its term of being to systems within its environment. Autonomy can never be solitary: in second-order cybernetics, autonomy is rethought as operational self-reference.

In brief, neocybernetics shifts the emphasis of observation and description from subject to system. One form of the neocybernetic turn is a shift of interest from the identities of subjects to the networks of connections among systems and environments. The humanist project that unified perception and communication in one subject, shored up against all odds by the first cybernetics, is now observed as an amalgamation or structural coupling of multiple observing systems. With this move the noumenal unity of the humanist subject gives way to a differential observation of the relations of living and nonliving systems and their environments, such as human and nonhuman bodies and societies.

Emergence and Closure

Emergence and Embodiment reflects on the legacy and continued, even renewed, value of neocybernetics in a world characterized by hyperacceleration in the sciences and technologies of emergence. In *We Have Never Been Modern*, Bruno Latour states that "we are going to have to slow down, reorient and regulate the proliferation of monsters by representing their existence officially."[15] In *Microcosmos*, Lynn Margulis and Dorion Sagan sound a similar note regarding humanity at large in its geobiological context: "The reality and recurrence of symbiosis in evolution suggests that we are still in an invasive, 'parasitic' stage and we must slow down, share, and reunite ourselves with other beings if we are to achieve evolutionary longevity."[16] Summarizing this broad neocybernetic consensus, Dirk Baecker writes that "One of the most important aspects of systems theoretical thinking is to proceed slowly, to look at things again, and to take the time to spell them out. . . . We should not jump, as systems do, from one event to the next simply to show that we can do so. Rather, we should look back at each instance, again as systems do, to see how we effected the last jump."[17]

Following these complementary invocations for a "slowing down" of technoscientific hybrids, of ecological depredations, and of systems-theoretical theorizations, we acknowledge a similar need for a slowing down—in our case of everything that has recently come together under the rubric of the "posthuman"—for the purpose of careful neocybernetic consideration.[18] By now already a cultural cliché lacking definitional consensus, the posthuman has been wielded to encompass everything from contemporary theorizing and cutting-edge cultural history; to work in nontraditional sciences like nonlinear dynamics, robotics, artificial life; and indeed to the science of emergence that has been dubbed (by its most ambitious proponent, Stephen Wolfram) "a new kind of science." By means of a performative polarization deploying classic binary logic, facile versions of posthumanism reproduce the human as the very "other"—the much despised and easily criticized figure of a unified and fully autonomous human subject—whose devalorization is to give them teeth.

As we see it, the human has always been a *for-itself* complexly imbricated with the environment, and it is precisely a recognition of this complexity that informs the historical moment of second-order cybernetics, as well as its continuation in what we are calling neocybernetics. Central to the priority we want to claim for neocybernetics is the concept of autonomy as double closure or, as Heinz von Foerster puts it, the regulation of regulation. In stark contradistinction to any naive conception of autonomy as the absolute self-sufficiency of a substantial subject, this concept demarcates the paradoxical reality that environmental entanglement correlates with organismic (or systemic) self-regulation. Thus a system is open to its environment in proportion to the complexity of its closure.

That this equation remains in force even—and indeed must remain in force especially—in the face of today's massive incursions of technics into the domain of the living is one of the central burdens of our volume. For if the human has

always been posthuman in the sense that it has always involved an exteriorization or evolution by means other than life (as the work of André Leroi-Gourhan
and, more recently, Bernard Stiegler has shown), the massive contemporary acceleration in processes of posthumanization poses the prospect of a qualitative
shift in the economy between autonomy and environmental entanglement.[19]
Whether this shift entails the abandonment of autonomy as regulation of regulation is a crucial question facing cultural theorists today.

One eloquent position here—that of Katherine Hayles in *My Mother Was a
Computer*—contends that recent technosciences of emergence and the model
of the computational universe they presuppose have definitively marked the
historical limits, indeed the eclipse, of the cybernetic tradition:

> Even the most insightful and reflective of the cyberneticians stopped short of
> seeing that reflexivity could do more than turn back on itself to create auto-
> poietic systems that continually produce and reproduce their organization.
> Heinz von Foerster's classic work *Observing Systems* shows him coming to
> the threshold of a crucial insight and yet not quite grasping it: the realization
> that reflexivity could become a spiral rather than a circle, resulting in dy-
> namic hierarchies of emergent behaviors. By the time scientists began to use
> this idea as the basis for new kinds of technologies, cybernetics had already
> lost its utopian gloss, and new fields would go by the names of artificial life,
> complexity theory, and cellular automata.[20]

As we see it, however, this evaluation seriously underestimates the force of von
Foerster's account of how recursive processes generate emergent complexity.
At the core of Hayles's claim is a conviction that the role of recursion as understood by second-order cybernetics simply cannot account for the processes of
emergence that are popping up everywhere in our world, whether one consults
the computational model developed by Stephen Wolfram in *A New Kind of Science* or the totalizing picture of escalating onto-genesis promoted by Howard
Morowitz in *The Emergence of Everything*. The "crucial question" of Hayles's
book is precisely how the "new kind of science" that informs what she calls the
"Regime of Computation" can "serve to deepen our understanding of what it
means to be in the world rather than apart from it, co-maker rather than domi-
nator, participants in the complex dynamics that connect 'what we make' and
'what (we think) we are.' "[21]

While we concur with this desideratum and consider it central to what we
are calling neocybernetics, we simply cannot endorse this position. For Hayles,
developing an understanding of our constitutive worldliness requires a trans-
gression or dismissal of the boundaries separating any system from its environ-

ment: "Boundaries of all kinds have become permeable to the supposed other. Code permeates language and is permeated by it; electronic text permeates print; computational processes permeate biological organisms; intelligent machines permeate flesh. Rather than attempt to police these boundaries, we should strive to understand the materially specific ways in which flows across borders create complex dynamics of intermediation."[22] In our view, these formulations are simply too vague. It is not at all clear what exactly such "permeation" might amount to, given the very different operational fusions being asserted. In her zeal to leave closure in the dust, Hayles simply glosses over the very differentiations from which systems are generated in the first place. This move short-circuits the machinery of emergence before emergence even gets started.

Neocybernetics contends that it is precisely the injunction against such flows of *information* across the boundary demarcating an autopoietic or self-referential system from its environment that drives the theory's crucial insights into the operations of self-referential and recursive forms. It is not a matter of "policing" operational boundaries: not only are they self-producing and self-maintaining, but they are the condition of the possibility of systemic functions in the first place. The environment can perturb living, psychic, and social systems but cannot *operationally in-form* them. More simply put, environmental stimuli can trigger systems to restructure themselves but cannot directly or causally impact their function. We can say, then, that systems' observations of their environment are internally and autonomously constructed by their own ongoing self-productions. In other words, to maintain their autopoiesis, (self-referential) systems must remain operationally (or organizationally) closed to information from the environment. On that basis, they can construct their interactions with their environment *as* information. Luhmann writes with regard to the operation of communication in social systems: "A systems-theoretical approach emphasizes the *emergence of communication* itself. Nothing is transferred."[23]

To forestall a misunderstanding that has dogged second-order systems theory since its inception, we need to insist upon the specificity of neocybernetics' complex, nuanced, and paradoxical understanding of the concept of *closure*. Once the paradigm shift is made from the physical to the life sciences, the order-from-noise principle in self-organizing systems gives way to the openness-from-closure principle in autopoietic systems. To understand the stakes of this development, one must bring into play the fundamental distinction between thermodynamic and autopoietic principles. Thermodynamically, a system is *either* open *or* closed to energic exchange with its environment; by contrast, autopoietic systems are *both* environmentally open to energic exchange *and* operationally closed to informatic transfer. According to this understanding, operational closure—far from being simply opposed to openness—is in fact the precondition for openness, which is to say for any cognitive capacity whatsoever. As a number of contributors demonstrate, this generalized correlation of closure with cognition informs Varela's development of neocybernetics—specifically his development of the openness-from-closure principle—from its initial theorization in relation to autopoietic systems to the meta-autopoietic assemblages that, arguably, characterize contemporary society.

In their various characterizations of autopoiesis, Maturana and Varela correlate organizational closure with interactional openness: it is an organism's (or system's) self-perpetuation that allows it to be structurally coupled to the environment. In this volume, Evan Thompson restates this core neocybernetic insight from the perspective of Varela's later work bridging life and mind, neuroscience and phenomenology, through the concept of autopoiesis: "The self-transcending movement of life is none other than metabolism, and metabolism is none other than the biochemical instantiation of the autopoietic organization. That organization must remain invariant—otherwise the organism dies—but the only way autopoiesis can stay in place is through the incessant material flux of metabolism. In other words, the operational *closure* of autopoiesis demands that the organism be an *open system*."[24] This nuanced concept of closure also informs Luhmann's remark that with second-order systems theory, "The (subsequently classical) distinction between 'closed' and 'open' systems," as that was previously defined in regard to allopoietic physical and mechanical systems under strictly thermodynamic regimes, "is replaced by the question of *how self-referential closure can create openness*."[25] Arguments that assume closure to be the simple binary opposite of openness fall short of the letter and complexity of neocybernetic conceptualization.

Put another way, in order for a system to perpetuate itself, it must maintain its capacity to reduce environmental complexity, which is to say to process it not as direct input but as perturbation catalyzing (internal) structural change. As von Foerster's "postulate of cognitive homeostasis" has it (and this would certainly hold for autopoietic systems in general), "The nervous system is organized (or organizes itself) so that it computes a stable reality."[26] The challenge we propose to take on here is precisely the one advanced by Hayles in the name of the contemporary technosciences of emergence. To Hayles's claim that neocybernetics cannot embrace the complexity of contemporary emergences and their permeation of systems boundaries, we reply that these processes *can be understood through the correlation of systemic closure and openness*. What is needed is a generalization of the openness-from-closure principle that is capable of addressing the full complexity of contemporary systems operations and environmental couplings. To develop such a generalization, we propose to explore how various facets of neocybernetics—running the gamut from Varela's work on living systems to Luhmann's account of communicational autopoiesis—deploy recursivity to underwrite emergence. Our efforts here are loosely guided by the following postulates. They are intended as initial steps toward specifying what we mean by neocybernetic emergence and toward generalizing its extension:

1. Neocybernetics requires a recognition that there are only two orders of cybernetics or, alternatively, that the shift from a first-order to a second-order cybernetics marks the passage to a general form of recursivity that *can* (contra Hayles) spiral outwards and thereby create the new at successively higher levels. While such a requirement inheres in all the neocybernetic accounts explored in our volume, it finds exemplary expression in von Foerster's claim that second-order cybernetics is a "cybernetics of cybernetics" and that a "third- [or higher-] order cybernetics . . . would not create anything new, because by ascending into 'second-order,' as Aristotle would say, one has stepped into the circle that closes upon itself."[27]

2. Neocybernetics facilitates a concept of emergence that differs in at least one fundamental way from the concept of emergence central to contemporary technosciences and the regime of computation. Whereas the latter understands emergence as a movement from the simple to the complex (cf. Wolfram's maxim: from simple rules, complex behavior), neocybernetics views it as a movement from the chaotically complex to the manageably complex.[28] In line with what Luhmann calls *decomplexification*, it is a given that any particular system that emerges within an environment is necessarily less complex than that environment (since the latter will always contain *many other* systems). Indeed, one of the capital advantages of the concept of the self-referential system (as against the notion of the subject) is its delineation of such a system's capacity to manage environmental complexity and indeed to derive its identity and its autopoiesis from its continual need to reduce the complexity of the environment by processing it through systemic constraints. Our endeavor here will be to produce viable accounts of emergence that meet the terms of Hayles's objection—and the body of research upon which it draws—by showing how neocybernetics can in fact account for the interplay of complexification and decomplexification in systems that do more than simply maintain their thermal homeostasis or basal autopoiesis. It is our conviction, moreover, that such an account is precisely what lies at the heart of neocybernetics and that what differentiates it from recent technosciences of emergence and computational

accounts of complexity is precisely its more fine-scaled and dynamic account of operational closure.

Here neocybernetics can endorse the objection raised by Ray Kurzweil against Wolfram's new science—to wit, that it explores the emergence of complex patterns at a first level of complexity, leaving aside the crucial issue of how these patterns self-organize to create higher levels of complexity.[29] And, indeed, neocybernetics lends philosophical substance to this objection, since the shift from first- to second-order cybernetics—from cybernetics to neocybernetics—is precisely what renders recursivity capable of self-organization and formal evolution. Once again, it is von Foerster who makes explicit the link between neocybernetics and the commitment to the motif of closure: "The essential contribution of cybernetics to epistemology is the ability to change an open system into a closed system, especially as regards the closing of a linear, open, infinite causal nexus into closed, finite, circular causality."[30]

This shift from an equivocal concept of openness to an operational concept of openness-from-closure underwrites a related shift from a representationalist to a constructivist epistemology and ontology. In this way, neocybernetics can address the wavering between two senses of emergence—epistemological and ontological emergence—that has dogged the computational model of emergence.[31] Wolfram can claim only to furnish a model of epistemological emergence, even if (or when) he wants to claim more than that; that is why the principle of "computational equivalence" (which states that a computer simulation of a complex process must be as complex as the process itself) furnishes the most powerful argument for his model of emergence. By contrast, neocybernetics, precisely because it invests in circular recursivity, seeks to develop a mechanism for explaining what Hayles (glossing Morowitz) calls "dynamical hierarchical emergences," which is to say "how one dynamical level enchains and emerges from the next lower one through their intersecting dynamics."[32] Here the extensive similarities between neocybernetics (cybernetics of cybernetics) and Morowitz's fourth (and to date final) stage of emergence (mind contemplating mind) are telling, since recursivity in both cases forms a powerful vehicle for the reduction of complexity that fuels emergence at the higher level. As specified through recursivity, the pruning algorithms that allow selection of probable conditions from a "transcomputable" possibility-space function in a very similar manner to *reentry*, the neocybernetic mechanism specified by George Spencer-Brown's *Laws of Form*, the recursive introduction of a system-environment distinction into the system itself (for more on Spencer-Brown's concept of *reentry,* see the essays by Clarke and Schiltz in this volume). Following in the wake of these similarities and also of the technosciences of emergence that are undeniably at work in our world today, the task facing neocybernetics—and our various explorations in this volume—can be specified to be that of showing how the ineluctable self-differentiation of a system that must maintain its autonomy over time can yield the emergence of the new, which is to say how it can yield emergence understood in its current usage as the appearance at the system's global level of properties that do not exist at the local level of a system's components.

Another way to understand this specification returns us to our claim concerning the specificity of the neocybernetic concept of emergence—namely, that in contrast to the technosciences of emergence, it proceeds not (like some latter-day Herbert Spencer) from the simple to the complex, but rather by way of system-specific and system-internal reductions of hypercomplexity to ordered complexity. This is the meaning of von Foerster's statement that it is we who invent the environment that we perceive and of similar claims appearing in Varela's conception of the "surplus of significance" and in Luhmann's account of contingent selectivity. Indeed, these claims all instance the operation of an epistemological constructivism that is common to all of the contributions in this volume, including those, like Mark Hansen's and Ira Livingston's, that manifest some doubts about Luhmann's strong constructivism. For these critics—and here is what differentiates their accounts of emergence from the more positivist technoscientific account of Hayles—the insistence on the value, indeed the necessity, of constructivism becomes all the more imperative in the wake of the unprecedented technological complexification of the environment that coincides with the massive dissemination of computational systems throughout society.

Indeed, both Livingston and Hansen turn to constructivism as a helpful, if not ultimately sufficient, resource for reconceptualizing human agency as technically distributed agency. Even if their contributions evince a belief that contemporary environmental complexity has largely outstripped the capacity of systems to reduce it, their conceptualization of hybrid forms of agency—encompassing humans and technico-environmental processes—continues to invest in the minimal Luhmannian commitment that shores up all the configurations of neocybernetics collected here: the idea that selection is key to instituting difference into what would otherwise remain undifferentiated chaos. For these critics, it is the case that environmental processes have independent agency in technically distributed cognitive processes, but this reality does not in any way obviate the value of partial and provisional closures—closures that cut across system-environment boundaries—for facilitating observation of technically distributed cognitive operations of "system-environment hybrids."

Distinction and Assemblage

At a more general level, the epistemological constructivism of neocybernetics provides new frames of observation disposing one to mark system/environment distinctions more rigorously. This is an analytical option like any other, but arguably it is one with real purchase on the ineluctable self-reference entailed by and embedded within all descriptions. The self-evident proposition at the self-referential origin of systems theory as a scientific discourse is that "there are systems," followed immediately by its corollary: "There are self-referential systems." When Luhmann goes on to indicate that all the heterogeneous assemblages of biotic and metabiotic systems and their environments are just as self-referential as the "self-reflective subject" of Western metaphysics that privileged itself on its supposedly unique possession of reflexivity, the neocybernetic description of systems cuts across the grain of classical logic. Whereas the term "reflexivity" retailed by Hayles, Morowitz, and others retains the subjectivist connotation of "reflection" in "the mirror of the mind," we have preferred the posthumanist terms *recursion* and *recursivity* to underscore that what is being named, from cells to servomechanisms to societies, are the recursive functions of operationally closed observing systems. However, even within the same neocybernetic model, there is an important difference in its theorization of biotic as opposed to metabiotic systems. As Luhmann points out, not all autopoietic systems use *meaning* as a way to "virtualize" the system/environment relationship. Living systems do not, "but for those that do"—psychic and social systems—"it is the *only* possibility."

Key tenets of neocybernetic logic thus run as follows: (1) there are systems; (2) observation is possible only on the part of an observing system; (3) systems are self-referential, and so, in their treatment of matters beyond themselves, paradoxical; and (4) that which is observed as paradoxical in our experience is not necessarily a cognitive aberration but is just as likely to be a necessary component of the possibility of any experience at all. If one presses this logic toward philosophical polemic, it might go like this: Many of the problems of modern social systems are exacerbated by faulty understandings of the real systematicities of things and thus, for all our vaunted rationality, by the subsequent unreal or invalid constructions (read: ideological mystifications) of causes and effects in the world. To better grasp the world that is their ostensible object, the organs of scientific knowledge must accommodate themselves to the recursive paradoxicalities of their own operation. In sum: our understanding of the world comes by way of an assessment of the world's impact on systems, which is to say on the very systems that give us cognitive purchase in the first place.

In part, then, the goal of this volume is to set aside a lot of the anachronistic semantics that have accreted to "systems" theory and replace them with more viable presuppositions. For instance, one can "rage against the system," but that rage will proceed on the basis of the momentary well-being of various affective, cognitive, and communicative *systems*. Neocybernetics underscores that there are operational horizons that put ultimate limits on the disorder of both physical and cognitive systems. At the same time, the evolution of systems feeds on anarchy—sometimes rage is an effective form of communicative irritation. This is just to say that neocybernetic systems theory at its best puts new eyes on and into the world, X-ray eyes that separate ostensibly unitary constructions—nominal identities still freighted with cultural capital but lacking in "substantial" ontology—into multiple systems references. Critics looking at neocybernetics who focus on this "negative" moment in its discourse tend to see it as a particularly soulless idiom of nihilistic deconstruction. What this critique clearly lacks is the conversion experience that "reverses the sign" of systems-theoretical distinction-making from subtractive to additive and that, like deconstruction, factors the play of supplementarity into one's habits of discursive comprehension. For what drives the work of thinkers like Luhmann and Varela, no less than the contributions that make up this volume, is an interest in bringing an ever-complexifying world into the framework of cognition. What distinguishes the neocybernetics approach from other contemporary cultural theoretical positions is an appreciation of the difficulties and complexities involved in doing precisely that. It is our hope that the essays in this volume will help readers to get that neocybernetic message—to appreciate the importance of systems-theoretical distinction-making—and will allow them to engage productively with second-order systems theory's rich potentials for further development.

Contents of *Emergence and Embodiment*

The interview with Heinz von Foerster that begins the volume took place less than a year before he died at the age of ninety-one. He had been interviewed many times by then, but unique to this interview are the answers he gives to questions about the further growth of his own neocybernetic brainchild. His remarks on the line of second-order cybernetic development, from Maturana and Varela's conception of autopoiesis to Luhmann's appropriation of it for social systems theory, put to rest any doubt one might have had regarding his opinion on the validity of social autopoiesis. Beyond that, the conversation documents the fabled vivacity of this great thinker, the cultural resources he drew on as an émigré from Wittgenstein's Vienna, and the "magical" nature of psychic systems in social communication, working out the resonances that create mutual understanding.

In "Heinz von Foerster's Demons: The Emergence of Second-Order Systems Theory," Bruce Clarke examines some of the prehistory of neocybernetics by reading von Foerster's key 1959 paper on self-organization through the hindsight of his early 1970s work that launched second-order cybernetics proper. Instructively for the culture of his particular scientific practice, von Foerster's discursive milieu is populated by old and new allegorical figures. Not one but two Maxwell's Demons bind thermodynamic to informatic self-organization in the 1959 paper, and his own creation, the Man with the Bowler Hat, links that earlier paper with "On Constructing a Reality" of 1973, by way of contrasting the singularity of metaphysical solipsism with the multiplicities of epistemological constructivism. Not only does it take multiple demons to conceptualize negentropy in informational systems, but it also takes the co-construction of at least two operationally closed observers to produce a reality: "Reality appears as a consistent reference frame for at least two observers." The concluding section of the essay unfolds this powerful statement from the 1959 paper as a prefiguration of the neocybernetic concept of *reentry*, by which the system/environment dyad recurs upon and ramifies within the system itself. In Luhmann's theory, the dyad of mutually closed psychic and social systems is capable of interpenetration and meaningful resonance just because both systems share this same paradigmatic operation, becoming "two observers" that construct out of their coupled autonomies the world as a reference frame for psychic and social realities.

Francisco Varela's "The Early Days of Autopoiesis" gives his account of the personal and cultural circumstances, the intellectual and academic milieus, within which the concept of autopoiesis was cultivated. Humberto Maturana and Heinz von Foerster play major roles in this narrative, as do other figures, including Jean Piaget, Ivan Illich, and Erich Fromm. Varela cites Wiener, Mc-Culloch, and von Foerster as "*the* pioneers of the *conjunction* of epistemological reflection, experimental research and mathematical modeling." Along with this background in cybernetic epistemology, Varela also stresses the importance of his philosophical readings from Husserl and Merleau-Ponty for the development of his scientific work. Throughout this engaging reminiscence of a turbulent and seminal period culminating in his self-exile from Chile in the aftermath of the assassination of Salvador Allende, Varela illuminates the paths that eventually led him "from autopoiesis to neurophenomenology."

As Varela has reminisced about the formative period of his own science, in "Life and Mind: From Autopoiesis to Neurophenomenology," philosopher Evan Thompson opens his essay, drawn from a talk given at the 2004 Varela Symposium in Paris, with a memoir of the circumstances of his earliest encounter with Cisco. Fifteen years after that meeting in 1977, Varela, Thompson, and Eleanor Rosch would publish *The Embodied Mind: Cognitive Science and Human Experience*, which introduced a general readership to the scientific work Varela et al. had developed by then at the conjunction of epistemological reflection and experimental research. The neocybernetic theme of operational recursion emerges here as "the 'fundamental circularity' of science and experience." Thompson follows out the lines that link neurophenomenological research to the "embodied mind" implicit in Maturana and Varela's initial formulations of autopoiesis and ultimately in Maturana's pre-autopoietic insights in "The Biology of Cognition," then traces forward Varela's own refinements of autopoietic cognition in his careful unfolding of embodied "sense-making" through a concatenation of recursive emergings, from life to self to world, and thence to "cognition, in the minimal sense of viable sensorimotor conduct." Emergence and embodiment dovetail when Thompson proposes terms for understanding Varela's late rapprochement with the notion of teleology, not as the source or goal of autopoietic organization, but as an emergent domain arising from the coupling of an autopoietic system to its enabling environment, its embodied world.

In "Beyond Autopoiesis: Inflections of Emergence and Politics in Francisco Varela," John Protevi also traces key turning points in Varela's work. Protevi focuses especially on the concept of emergence, which was always central for Varela, and on questions of politics, which operate at the margins of Varela's thought. He divides Varela's work into three periods—autopoiesis, enaction, and radical embodiment—each of which is marked by a guiding concept, a specific methodology, a research focus, an inflection in the notion of emergence, and a characteristic political question. Protevi investigates the implicit "political physiology" of Varela's work—that is, the formation of political states and politically constituted individuals and their intersection in political encounters. Protevi maintains that in each register of political physiology the emergence of systems should be thought in terms of the resolution of the differential relations of a dynamic field. Varela had to move "beyond autopoiesis," in Protevi's view, precisely to be able to thematize such dynamism, as the recursive structure of the autopoietic system inhibits the ability to conceive of dynamic change. In other words, for Varela autopoiesis is bound to synchronic emergence (part-whole relations), whereas enaction can account for synchronic and diachronic emergence (creation of novel organization), and radical embodiment can account for synchronic, diachronic, and transversal emergence (body-brain-environment loops). Protevi sees this latter, wider conceptual scheme as necessary to an understanding of political encounters in all their dimensions.

In line with Thompson and Protevi, in "System-Environment Hybrids" Mark Hansen also focuses on Varela's conceptualization of emergence with the coupling of autopoietic systems and embodied environmental worlds. His specific aim here is less to explicate the trajectory of Varela's thinking for itself than to position Varela—and specifically Varela's decoupling of autonomy (closure) from autopoiesis in his key 1979 text, *Principles of Biological Autonomy*—at the origin of a mode of conceiving contemporary cognitive agency as massively technically distributed. Varela's decoupling of autonomy from autopoiesis facilitates the deployment of closure at a higher level of inclusiveness and with a complex internal differentiation. Such a modification is necessary, Hansen argues, if we are to theorize the multiple and differentiated levels of autonomy that characterize what he calls "system-environment hybrids" (SEHs), complex, hybrid forms of embodied, cognitive enaction that involve human cognizers coupled with technically advanced environmental processes wielding their own agency. Drawing inspiration from Bruno Latour's description of contemporary human/nonhuman hybrids as "rather horrible melting pots," Hansen positions SEHs at cross-purposes to Luhmann's description of operationally closed systems functioning through effective decomplexification of the environment. According to Hansen, it is precisely such decomplexification that has become both highly problematic and atypical in today's technosphere, where we are continually acting together with cognitively sophisticated machines; in our technosphere, the agency and complexity of the environment simply cannot be reduced to a function of a system.

While Hansen follows Luhmann in maintaining that selection is key to instituting difference into what would otherwise remain undifferentiated chaos, he departs from Luhmann when he asks whether the institution of difference might rather cross over system-environment boundaries and thus underwrite hybrid forms of agency comprised of human beings and complex technological processes. In this endeavor, Hansen draws on the work of Katherine Hayles, Andy Clark, and Félix Guattari, all of whom argue for the need to complicate the concept of closure in light of the technically rich environments in which we live and act in our world today. To develop a strong account of environmental agency, Hansen turns to two French thinkers—political philosopher Cornelius Castoriadis and biophenomenologist Gilbert Simondon—whose work helps to expand the impact of Varela's decoupling of autonomy from autopoiesis. Combined with Varela's insistence on the integrity of the human and on continuity across divergent levels of being, Castoriadis's differentiation of levels of autonomy and his conception of radical creativity and Simondon's privileging of the agency of the environment (what he calls the "preindividual") in the operation of all processes of individuation furnish the tools necessary to theorize, in a broadly neocybernetic mode, the functioning of SEHS that emerge in the wake of the contemporary complexification of our technosphere. Rather than possessing institutional (autopoietic) closure that cuts across the human, today's SEHS are created and dynamically evolve through what Hansen calls "technical closures," provisional forms of closure facilitated by contingent conjunctions of humans and technologies.

One major goal of this volume is to work through the lingering controversies over the purchase and application of the concept of autopoiesis, without blurring the important differences staked out by the key theorists involved but also in hopes of illuminating the shared commitments that gather the wider discourse of neocybernetics together. Thus we have deliberately brought the work of Francisco Varela and Niklas Luhmann into direct contact, and we invite our readers to determine their own positions within the powerful neocybernetic force field they generate between them. In "Self-Organization and Autopoiesis," our excerpt from *Einführung in die Systemtheorie* (Introduction to systems theory), Luhmann introduces his reformulation of autopoiesis by distinguishing it from another, closely related but earlier, systems concept with which it is often confused, namely, self-organization. This concept arose in the heyday of first-order cybernetics but even then marked the turning of classical cybernetic interest toward the manifestly autonomous behaviors of biotic and metabiotic systems, relative to their mechanical and computational counterparts. In contrast, autopoiesis was a second-order cybernetic concept from its inception, marking the initial fulfillment of von Foerster's heuristic formulations of "recursive mechanisms in cognizing systems."

The fact that self-organization remains a fundamental concept in the contemporary sciences of emergence indicates a sidelining or dismissal of the second-order, autopoietic approach. We see it as a kind of first-order hangover of the atavistic desire to endow computational systems with the facilities of a body, a desire best epitomized in the development of artificial life research, and brought to powerful and ironic narrative realization in Richard Powers's *Galatea 2.2*. Luhmann formulates the distinction in concise terms: self-organization relates to autopoiesis as structures relate to systems. The important point is not that certain systems "are" self-organizing, but rather that because certain systems are self-referentially or operationally closed, their formation of internal structures can result only from processes of *self*-organization. As Luhmann puts it, "there is no importation of structures from elsewhere." To develop this point, Luhmann works through a series of notions—expectation, memory, and the determination of the observer—relating the structurality of autopoietic systems to their *temporality*. Systemic structures partake of the time of the system and have effect only in the present moment of its operations. This contingency is most concretely obvious in the life of a cell but is also enshrined in the truism one applies to psychic and social systems, minds, and relationships: *Use it or lose it*; an autopoietic system is always about the business, from moment to moment, of reconstructing its structures.

The continuation of the Luhmann excerpt takes up autopoiesis proper and includes some important reminiscences of his conversations with Maturana. For instance, Luhmann gives his version of the anecdote concerning the *praxis/poiesis* distinction that Varela related about Maturana in "Early Days." He also joins von Foerster in commenting on Maturana and Varela's hesitation to apply the concept of autopoiesis to the processes of social communication. It is our hope that the overlapping concerns of the readings assembled in this volume will help to bury this bone of contention and allow everyone concerned with the further development of neocybernetics to move on to more fruitful initiatives. What unites all of the essays assembled here is the concept of operational closure. As Luhmann explains, the recognition of operational closure "is connected to a break with the epistemology of the ontological tradition that supposed that something from the environment enters into the one who cognizes and that the environment is represented, mirrored, imitated, or simulated within a cognizing system. In this respect the radicalism of this innovation is hard to underestimate."

In "Space Is the Place: The *Laws of Form* and Social Systems," Michael Schiltz examines a key resource for neocybernetic innovation, George Spencer-Brown's *Laws of Form*, a work of overarching importance in Luhmann's later studies of the functional systems of society and their culmination in the—deliberately paradoxically formulated—"society according to society" (*Die Gesellschaft der Gesellschaft*). Unsurprisingly, Spencer-Brown's "calculus of indications" had previously captured Heinz von Foerster's attention. In his *Whole Earth Catalog* review first run in 1970, von Foerster enthusiastically proclaimed *Laws of Form* a book that "should be in the hands of all young people."[33] In 1975 Francisco Varela deemed it a "calculus for self-reference."[34] Yet for all its seminal qualities, *Laws of Form* remains a subject of contention, particularly on account of its dense, at times koan-like, vocabulary (for instance: "distinction is perfect continence"). In Schiltz's illuminating treatment, Spencer-Brown's calculus presents a protologic of distinctions, rehearsing the forms of any possible observation. From this angle, it derives its importance from its unusual realization and innovative expansion of topological awareness. It addresses something not often realized, the contingency of Euclidean space—in particular two-dimensional space—and demonstrates that by reconceiving the form of space, we may meaningfully and more easily conceive of forms that "reenter" their own space. For instance, distinctions written on a torus "can subvert (turn under) their boundaries, travel through the torus, and reenter the space they distinguish, turning up in their own forms."

This reconceptualization of space has wide-ranging consequences for epistemology. For example, it informs Luhmann's description of the autopoiesis of psychic and social systems, which must reenter the system/environment distinction into themselves in order to observe their environments. By stressing the operative nature of this process, Spencer-Brown thus presents the mathematical foundations underlying second-order cybernetics' insistence on constructivism and self-reference, or autology. Its long-standing appropriation of *Laws of Form* shows how fundamentally second-order cybernetics' view of the possibilities and conditions of knowledge differs from traditional epistemologies. Shifting from the world of things to the world of observations, self-reference comes full circle: "Our understanding of the world thus cannot reside in some form of discovery of its present appearance (out there, beyond observation), *but comes from remembering the conventions agreed to in order to bring it about.*"

Drawing on Luhmann's systems theory and Spencer-Brown's concept of form, Edgar Landgraf's "Improvisation: Form and Event—A Spencer-Brownian Calculation" theorizes the operational closure of the art system and its consequences for our understanding of the artist, the creative process, and the experiencing of art. The first part of the essay looks at improvisation historically and argues that the twentieth century's celebration of spontaneity and improvisation in art, as well as the emphasis put on performance and effect, are long-term consequences of aesthetic codes that became dominant in the late eighteenth century. These codes secured the reproduction and social autonomy of art but also challenged traditional notions of agency in art. The aesthetics of genius reacted to these challenges with paradoxical figurations of intentionality. In their place, Landgraf suggests that we understand the art-creating process as self-ordering, as a process that reduces the complexity and contingency it finds in its environment according to programs it devises for itself. Such descriptions of the creative process are able not only to theorize the autonomy and heterogeneity of artist, artwork, and art system, but also to account for the increased prominence of contingency and improvisation in modern art.

In the second part of his essay, Landgraf explores the neocybernetic shift from an ontological to an operational viewpoint, in order to account for the emphasis improvisation (and contemporary art in general) puts on performance and effect over depiction and meaning. In line with Schiltz's discussion of *Laws of Form* as a protologic, Landgraf shows how Spencer-Brown's form concept allows us to conceive of art in pre-representational terms: we can understand the "experience" of cognitive engagement with art without having to assume an interpretive stance toward the work, but also without having to subscribe to ontological notions of "materiality" or existentialistic definitions of the human body and our being-in-the-world. Instead, we can comprehend the artistic *event* as created by the multiple, conscious and subconscious, operations the psychic and nervous systems (learned to) perform when they observe, relate to, identify with, ignore, reflect on, and let themselves be surprised by the artistic irritations they find in their environment.

Linda Brigham's "Communication versus Communion in Modern Psychic Systems: Maturana, Luhmann, and Cognitive Neurology" continues the focus on perceptual systematics by assessing the relationship between modern social time and some peculiar instances of so-called temporal disruption in individual experience. Her essay explores the impact of the global, univocal time that has become increasingly necessary for the functioning of advanced technological society, a society in which performing appropriate actions at appropriate moments is critical. Global modernity demands the orchestration of a huge array of human activities and places a premium on the interpenetration of unambiguous linear temporal measures with psychic systems. In order to illuminate the implications of this interpenetration, Brigham explores three instances in which psychic systems hesitate or fail to articulate themselves linearly in time. The phenomenon of the phantom limb, the intrusive memories that often follow trauma, and the lingering sense of the dead that constitutes grief all comprise conditions in which the past is experienced as in some way present. Brigham's accounts of these temporal disruptions portray them as posing differing degrees of threat to linear temporal autobiography.

Brigham argues that these temporal disruptions of autobiography, particularly trauma and grief, introduce affective limits to the inroads of communication—in Luhmann's sense of the medium of social systems—on psychic systems and instead constitute the basis for affective contagion, a source of social cohesion in which the closure between psychic systems and social systems appears to be in some way breached. Indeed, it may be that in the context of affective contagion certain high-level autopoietic distinctions between system and environment are radically altered, and a range of salience emerges that is not generally at the command of psychic systems interpenetrated by modern social systems. This scenario suggests that Fredric Jameson's observation of the waning of affect in modernity has a systems-theoretical foundation.

Cary Wolfe's "Meaning as Event-Machine, or Systems Theory and 'The Reconstruction of Deconstruction'" returns to the neocybernetic disarticulation of psychic and social systems. Decisively confirming the breadth of systems theory's philosophical credentials, Wolfe's essay aligns the work of Niklas Luhmann and Jacques Derrida as both converge on the problem of meaning—its form, its systematicity, and its function. Wolfe argues that Derrida and Luhmann bring to bear on the question of meaning remarkably similar theoretical stances whose chief characteristics include difference, differentiation, distinction, and temporalized complexity. This convergence has been difficult for readers to grasp because Derrida and Luhmann approach the same theoretical terrain from opposite directions and with rather different purposes. Where Derrida assumes the entrenchment of an always already logocentric philosophical tradition that must be shown to deconstruct itself, in the process revealing the various forms of difference and complexity of meaning (Derrida's "writing") that such a text represses, Luhmann's functional analysis is concerned with how difference and complexity are adaptive *problems* for systems that need to continue their autopoiesis in the face of overwhelming environmental complexity. For Luhmann as for Derrida, systems use "codes" to reduce and process complexity, but for Luhmann as for Derrida (as in his concept of the *grammé* in *Of Grammatology*), the fundamental nature of those codes is self-referential paradox that cannot be overcome but only temporalized.

These shared theoretical commitments enable Derrida and Luhmann to *disarticulate* psychic and social systems, consciousness, and communication (a position long familiar to readers in Derrida's critique of the elevation of speech over writing and the auto-affection of the voice as presence in his early work), the better to specify the ways in which they do, and do not, interpenetrate. For both, communication is possible only as a form that transcends dependence upon perception and consciousness. As Derrida puts it, meaning understood as "writing" *comprehends* language in the more narrow sense. Both communication and consciousness, however, use the form of meaning, and the medium that allows their interpenetration is language, a second-order evolutionary achievement that, Luhmann writes, "transfers social complexity into psychic complexity" and one that becomes more and more powerful—more and more communicative—in the evolutionary drift from graphic and alphabetic writing to printing, which further decreases the reliance of communication upon perception.

Both Derrida and Luhmann, then, undertake two crucial disarticulations that make their work thoroughly anti-representationalist and resolutely posthumanist: on one hand, of psychic systems and consciousness from social systems and communication, and on the other, of language in the strict sense as a type of "symbolically generalized communications media" (Luhmann) from the more fundamental dynamics of meaning that comprehend it. For both thinkers, in other words, language may be human, but meaning is not, and this allows us to think the relations between human and nonhuman worlds (technical, social, animal, and biological) with a renewed appreciation and understanding for the henceforth virtual space that they co-constitute in their processes of making meaning.

In "Complex Visuality: The Radical Middleground," Ira Livingston provides our volume a coda from the side angle of cultural studies. Livingston examines a number of the claims this volume makes for neocybernetics by viewing them through the lens of contemporary visual culture. What, Livingston asks, can postmodern constructions of visuality "tell us about notions of emergence, complexity, and systematicity?" In line with Hansen's interrogation of the potential for "system-environment hybrids" to breach the system/environment distinction and Brigham's investigation of cognitive phenomena that appear to suspend notions of operational closure in psychic systems, Livingston seeks to delineate a "radical middleground" wherein the problematics of the contemporary visual field situate a continually emergent intermedial space between figure and ground, system and environment. Occupying his own discursive middleground both inside and outside the parameters of neocybernetic concepts, Livingston offers a performative critique of this volume's philosophical polemics. Sympathetically skeptical, Livingston's pointed inquiries clarify what is at stake in the volume's strong construction of neocybernetics' continuing relevance to current theoretical and cultural debates.

Emergence and Embodiment is a collective effort to update the historical legacy of second-order cybernetics. In order to understand today's hyperacceleration of technoscientific incursions into the human and in order to arrive at more highly articulated observations of the systemic situatedness of cognition, all of the contributors correlate epistemological closure with the phenomena of ontological emergence. In this respect, and despite their diversity, they forcefully testify that the latter cannot be understood independently of the former. The contemporary understanding that the human is and has always already been posthuman could not have emerged, and cannot be rendered productive, without the perspective afforded by neocybernetic recursion.

Notes

1. Varela, "Introduction: The Ages of Heinz von Foerster," xvii.
2. Ibid., xvi.
3. Ibid., xvii.
4. Wiener, *The Human Use of Human Beings*, 8.
5. Ibid., 1–2.
6. See Ashby, *An Introduction to Cybernetics*.
7. Bateson, "Cybernetic Explanation," in *Steps to an Ecology of Mind*, 401.
8. Ashby, *An Introduction to Cybernetics*, 1.
9. Wiener, *The Human Use of Human Beings*, 109.
10. Ibid., 110.
11. Luhmann, "The Medium of Art," 218.
12. Von Foerster, "On Constructing a Reality," 211.
13. Ibid.
14. See for instance Atlan, "Hierarchical Self-Organization in Living Systems"; Maturana and Varela, *Autopoiesis and Cognition* and *The Tree of Knowledge*; Margulis, "Big Trouble in Biology" and *Symbiosis in Cell Evolution*; and Oyama, *The Ontogeny of Information*.
15. Latour, *We Have Never Been Modern*, 12.
16. Margulis and Sagan, *Microcosmos*, 196.
17. Baecker, "Why Systems?" 70.
18. For an extended rehearsal of the "neocybernetic posthuman," see Clarke, *Posthuman Metamorphosis*.
19. Leroi-Gourhan, *Gesture and Speech*; Stiegler, *Technics and Time*, vol. 1: *The Fault of Epimetheus*.
20. Hayles, *My Mother Was a Computer*, 279–80.
21. Ibid., 280.
22. Ibid.
23. Luhmann, "What Is Communication?" 160; original emphasis.
24. For the full development of this line of thought, see Evan Thompson, *Mind in Life*.
25. Luhmann, *Social Systems*, 9; our emphasis.
26. Von Foerster, "On Constructing a Reality," 225.
27. Von Foerster, "Ethics and Second-Order Cybernetics," 301.
28. See Wolfram, *A New Kind of Science*.
29. See Kurzweil, "Reflections on Stephen Wolfram's *A New Kind of Science*."
30. Von Foerster, "Cybernetics of Epistemology," 230.
31. Silberstein and McGeever, "The Search for Ontological Emergence."
32. Hayles, *My Mother Was a Computer*, 30.
33. Von Foerster, "*Laws of Form*."
34. Varela, "A Calculus for Self-Reference."

Interview with Heinz von Foerster

July 20, 2001, Pescadero, California

INTERVIEWER: BRUCE CLARKE

Family in Vienna

Bruce Clarke: I wanted to ask you about your very unique style, your playful way of putting professional papers together. Did you always write that way?

Heinz von Foerster: I think my answer is that I'm from Vienna. At the time I was born, at the turn of the century, Vienna was so multicultural—fabulous in medicine, in architecture, in art and painting and drawing. Ernst Mach and people like that came from Vienna. My family belonged to this whirlwind of people. My father was an architect with the electric industry but had lots of friends in mathematics and physics. My mother came from an artistic family. Her people were dancers, painters, sculptors, poets, and I was a little kid tossed into this bunch of different people who met at the home of my parents. My grandmother kept a kind of salon where people from different universes met. The actress Eleanor Duse came to the house.

What was your grandmother's name?

Marie Lang. She published the first European journal on women's liberation, and she was therefore known all over Europe.

Did they have the suffrage issue—the women's suffrage movement that began in the nineteenth century?

They were not directly members of it. My grandmother founded something which came from the British suffrage movement . . . the settlements, for poor people. And my mother and some of her friends founded one of those settlements in Vienna. So when I was a little boy, staying with my grandmother, sitting under her gigantic desk with big legs, I had my little chair there, and then all the ladies argued about philosophy, women's labor, and the rights of women. So as a kid I was familiar with the political and the cultural problems which arise in a society.

Uncle Ludwig

In many cases people are victims of semantics. They are not aware of what they are saying. My role is to be a semantic cleaning boy, who comes out with a big broom.[1]

That was Wittgenstein's point.

Exactly.

To sweep the pointless arguments off the stage.

Exactly. You see, I was a Wittgenstein victim when I was nineteen or twenty. I think I knew the *Tractatus* by heart. . . .

Now are you related to Wittgenstein?

Yes, I am related distantly.

You have a cousin named Wittgenstein. . . .

Yes, my grandmother married twice. And from her first marriage there are children that are related directly to Wittgenstein. I knew Uncle Ludwig when I was about eleven or twelve. I think I can even localize the time when I met him. My mother was a very good friend of his sister, Margarethe. They were very, very close. I was with her visiting Aunt Margarethe, and a young man came in. He asked me, "Heinz, what would you like to become when you are grown up?" Now I had just passed a very crucial examination. I wanted to get out of grade school and go into the gymnasium. Wittgenstein asked me what I would like to become, and I said I would like to be a *Naturforscher* [scientist]. To me that was a combination of Fridtjov Nansen and Madame Curie. He said, but then you have to know a lot. I said, but I know a lot; I just passed the examination. And then he said, you know a lot, but you don't know—how right you are. I said, what? I don't know how right I am? This was my first encounter with Uncle Ludwig.

Later on I became deeply involved in the *Tractatus*. And there was a nephew, a real nephew of Ludwig's by the name of Karl, who also knew the *Tractatus* by heart: "Heinz, can you tell me proposition 6.24?" I said, "Of course, that's an easy proposition." "Yes, but Heinz what about 7.1?" . . . I was really a pest: "Apologies, ladies and gentlemen; what you are saying is all wrong. According to proposition 2.7 in the *Tractatus . . . that is the case."* They said, "Poor Heinz, what can we do with you?"

Luhmann and Maturana

Were some of your papers driven by the invitations you received?

In most cases it was the consequence of an invitation, either to write a paper or to give a lecture. I tried to strike a balance, to give a paper people should enjoy. I don't want to talk gibberish that nobody understands. Who are these people and what are they interested in and why did they invite me? This is what I ask myself first, and then I sit down and say, what can I tell them?

As it was with the Luhmann thing, you see.[2] They were all academics. I wanted to do two things. Number one, even if you are not a sociologist, you can say very tough things about sociology which are not easy to digest and which are not being observed, even by the sociologists themselves. The other thing is that you can make them laugh. So I had the flower bouquet—the mathematics for afterward.

Is it easy to get a magical bouquet?

Of course, if you are a professional magician, you can produce them. It is no problem. I was in fact looking forward to that moment: to see what will be on these professors' faces when I produce that flower bouquet.

Now were these Luhmann's colleagues?

I think it was the whole faculty of the University of Bielefeld and the members of the research organization that celebrated Luhmann's retirement. So it was mostly high academia collected there.

Did you have much interaction with Luhmann over the years?

No, not at all. Luhmann himself corresponded a lot with me. He visited me; he was here himself, sitting on this chair; and he was very interested in me because he knew Heinz understands autopoiesis. And he wanted to articulate autopoiesis for social theory. And I always told him, "Be very careful; you will step on Maturana's toes. The most sensitive things, maybe there are already some swellings on his toes." But I know Maturana very well. In fact I was present when Varela and Maturana invented the notion of autopoiesis and wrote their first paper. When I left Chile, I took that paper, which was not really completely written, and with one of my students we finished it up so that the

Spanglish became English. I even wrote into that paper, "We thank Heinz von Foerster very much for the editorial help."[3]

Well, that's only right! Now I came into systems theory a couple of years ago. I did a seminar at Cornell last summer with David Wellbery, and we learned a lot of Luhmann. And then I started talking Luhmann with some of my literature and science friends, and they're Maturana and Varela fans.

That's the two directions.

And I'm trying to understand why—what is their problem with Luhmann?

Their problem with Luhmann is they [Maturana and Varela] do not want to have autopoiesis applied to social theory. This is just a personal idiosyncrasy. They would like to keep autopoiesis solely for biological discussion or biological research.

But Maturana and Varela seem also highly socially motivated and make their own ethical statements on the basis of autopoiesis. . . .

Particularly Maturana. He doesn't want his ideas even being mentioned by someone else.

Well, the cat is out of the bag now.

Changing Social Theory

I flew to the German cybernetics society meeting in Nuremberg.[4] So I said, going over I can think or dream a paper up during the flight. With eighteen hours between San Francisco and Frankfurt, it'll be easy for me to write up that paper. When I was sitting in the airplane, I realized I had not used my German for twenty years, and I had no idea what to do about it. So I was writing that paper for the German cybernetics society the whole way. When I arrived there, I took my scribbled things out and went to the conference. It was in the Meistersinger Halle, a very large conference room with about two thousand people or something like that. So I started to read and I couldn't; at the moment I stood on the lectern, they switched on four thousand watts of lights because it was of course a televised affair. So I was completely blind to the audience. Under the tremendous four-kilowatt lamps, sweat was running down my forehead, the heat was unbearable, and then I tried to invent my paper. But apparently I succeeded. When I was through, I had to get into fresh air; I couldn't stay in that room anymore. So I walked out, sitting on the sidewalk, and a gentleman came running up to me. "Do you know what they were saying?" I said, "No, I couldn't listen to myself; it was too hot." "They said, 'You've changed social theory!'" I said, "I hope to the better." He went on, "You have to come to conferences." The firm Siemens had organized a program for establishing all the foreign workers, whether from Turkey, from Romania, who came to work in Germany.

The Gastarbeiter.

Pools for the children, entertainment for the grownups, etc. etc. He said, "You've changed social theory. You have to tell them what to do about the Turks in Germany." I said, "Very easy—import Turkish girls."

"Objects: Tokens for (Eigen-)Behaviors"

Immediately the next day I took the train to Geneva, there to present this story-object, which was to celebrate Piaget's eightieth birthday.[5] So I presented that story. It produced the greatest uproar of antagonism I have ever received. When I finished with "thank you very much," I only could hear "Boo!" They were really jumping at me. "Whatever mathematics you have presented here has nothing to do with mathematics; in fact it has nothing to do with anything! I don't know what you are doing here! Where did you come from, from Mars?"

Well, I was in a very good mood overall. . . . Whenever I got some of the broadsides shot against me, I had a funny answer. So everybody laughed and nobody was angry. It was a very entertaining hour of argumentation, where people said to me, "This is utter nonsense!" And I said, "Why do you say utter? Why don't you just say nonsense?"

I confess I found it difficult going.

This is a very difficult paper.

The first time I read it, I really had no idea. . . .

It's utterly incomprehensible.

But I have come back to it recently, and this is now after studying systems theory and reading Luhmann and also reading some Varela.

It was Varela who said this is my most important paper.

I think I see why he would say that, because it compresses the argument about circularity almost as far down as one could do.

Precisely, yes.

So let me try to describe your argument.

Yes, wonderful. I would be delighted.

Well, if the nervous system is organizationally closed, and this is the case for you and for me and for all nervous systems. . . .

Exactly.

Then how is it that we can agree on the world outside of us? Since we're both inventing it for ourselves all the time. And whereas in a traditional approach, one would say the world is full of objects and they present themselves to us and we simply are aware of their existence because our nervous systems represent them to us or give the objects to us, then there's no problem. But if you're going to be rigorous about a constructivist epistemology, then you should not talk about objects. Because we don't know them.

Exactly.

So it's an effort simply to shift the vocabulary of our discussion, and so let us say that objects present "tokens for eigenbehaviors," which we can *establish.*

I think you understood it very well. I've never gotten such a good report about that paper! I do think there is one point missing in your story, but it is only missing in your story and not in your knowledge. This is that we are both in our world, both in each other's world. You are in mine, and I am in your world; therefore we establish our eigenbehavior for each other. And we may not agree, but we are caught in the same loop.

That's the two ourobori together.[6]

If I do that with my eyeglasses—I never poke them in my eyes, I put them properly on the nose, I find my pocket where the eyeglasses go. This is a very stable behavior. Piaget published a crucial paper making the key point that we can understand things only by handling them, by moving them, by moving our own body.

Right. Or we realize that our nervous system is operating in terms of these feedback loops, and so you need the confirmation of the two.

Exactly. There's a circularity. Piaget's central point was that you need the motorium to understand the sensorium. If you only look, you will not understand. You have to touch, you have to move, and then you will understand.

Or grasp—that's a good metaphorical statement. You have to grasp things in order to grasp them. . . . There are easier papers than that!

The Magician

Did you get the paper on Luhmann?

Yes. . . . "How Recursive Is Communication?"

Yes, it's a little bit queer. My problem was that all the great German professori were sitting in this gigantic room, all with beards, and they were such caricatures I couldn't believe my eyes. There was not a single smile to be seen. And I had to do something to produce a smile from one of these guys. I used to be a magician when I was a kid.

Oh, I see.

And so I pulled out a bouquet of flowers.[7]

Well, that makes sense. For instance, in your paper "On Self-Organizing Systems and Their Environments," you begin with two paradoxes.

Exactly.

Do you think that is the magician touch?

Exactly. And the interesting thing is that the magician is doing just the opposite of what most people think—hiding something. No, the magician is making things so clear that everybody can see what is going on. And that is the miracle. You must let them see the miracle, making it so convincing that absolutely nothing is hidden, nothing is under the table, everything is on the table, and that makes the whole thing very magical.

Notes

1. As evidenced, for instance, in von Foerster's essay "Molecular Ethology."
2. "The Luhmann thing": "Kommunikation und Gesellschaft: Autorenkolloquium für Niklas Luhmann," a one-day program at the University of Bielefeld on February 5, 1993, organized by Dirk Baecker and Peter Weingart for Luhmann's sixty-fifth birthday and retirement, at which von Foerster gave the first presentation, the essay cited in note 7 below.
3. Varela, Maturana, and Uribe, "Autopoiesis." The article contains an acknowledgment that reads as follows: "The authors wish to express their gratitude to the members of the Biological Computer Laboratory of the University of Illinois, Urbana, particularly

to Richard Howe, Heinz Von Foerster, Paul E. Weston and Kenneth L. Wilson, for their continuous encouragement, discussions, and help in clarifying and sharpening the presentation of our notions" (193).

4. Von Foerster refers here to the lecture he gave on March 28, 1973, at the Fifth Congress of the German Society of Cybernetics, later published as "Cybernetics of Epistemology." In this interview he conflates that trip abroad with the one that took him to Geneva three years later (see next note).

5. Von Foerster, "Objects." A footnote to the title reads, "This contribution was originally prepared for and presented at the University of Geneva on June 29, 1976, on occasion of Jean Piaget's 80th birthday" (261).

6. A circle of two snakes biting each other's tails, an illustration from "Objects," in *Understanding Understanding*, 267.

7. "Editors's [*sic*] note: Heinz von Foerster hands Niklas Luhmann copies of three articles specially bound for this occasion; as he does so, a bouquet of flowers appears magically out of thin air and Niklas Luhmann thanks him" (von Foerster, "For Niklas Luhmann" in *Understanding Understanding*, 308). In the second Luhmann folder at the Heinz von Foerster Archiv, Institut für Zeitgeschichte, Unversität Wein, in addition to the information about the Autorenkolloquium cited in note 1 above, there is a receipt for $33.59, dated January 4, 1993, paid to the House of Magic, San Francisco, California, presumably for the magical bouquet. Thanks to Albert Müller of the Heinz von Foerster Archiv for his hospitality.

Heinz von Foerster's Demons

The Emergence of Second-Order Systems Theory

BRUCE CLARKE

At its inception the discourse of cybernetics centered on the cluster of topics given by the initial title of the famed Macy Conferences, ten of which occurred between 1946 through 1953, "Circular Causality and Feedback Mechanisms in Biological and Social Systems." The interdisciplinary group here assembled brought together philosophically minded scientific polymaths and pioneers of electronic computation and information theory, such as Warren McCulloch, Norbert Wiener, John von Neumann, and Claude Shannon, with anthropological social scientists such as Margaret Mead and Gregory Bateson. A student of the Vienna circle trained in mathematics, physics, and electrical engineering, Heinz von Foerster in 1949 landed newly arrived in the United States and somewhat miraculously in the midst of this uncommon aggregation, in the middle of its run, and, despite minimal proficiency in English, was appointed (by McCulloch) the secretary of the Macy Conference proceedings. A year later and in the influential wake of Norbert Wiener's 1948 book, *Cybernetics, or Control and Communication in the Animal and Machine*, von Foerster suggested changing the name of the Macy Conferences to simply "Cybernetics," and his suggestion was adopted.[1]

The Macy Conferences represent the high point of the first interdisciplinary synthesis through which cybernetics came forward as a metadiscipline, bringing physical, mathematical, and engineering concepts of entropy, information, and feedback toward an integrated study of complex mechanical, computational, biological, psychic, and social systems. However, in the years after the Macy Conferences closed up shop, this cybernetic synthesis gradually splintered into noncommunicating specializations. Broadly considered, it diverged sharply back into subject/object dichotomy and Cartesian dualism—what could be called "hard" and "soft" camps. The former monopolized its resources, hoarded its grants, and redirected the mathematical and engineering sides of cybernetics toward Artificial Intelligence (AI), robotics, computer science, and command-control-communications technologies. To its credit, this is why you now have a computer on your desk and an iPhone in your pocket. The latter camp, often loosely identified with the work of Gregory Bateson, gradually gathered up the cognitive and philosophical insights of cybernetics toward matters of managerial and social systems, psychotherapy, and epistemology. Few persons besides von Foerster could be said to have had a foot in both camps, and no other vision of a holistic cybernetics was forthcoming to split the difference between them. Instead, the abandoned middle ground was eventually filled up with a multifarious cultural mythology centered on celebratory and cautionary images of the cyborg, a theoretical figure built up from variously real and imaginary mergers of biological bodies and electronic brains.

Twenty years after the lapse of the Macy Conferences, however, von Foerster would consolidate these alternative cybernetic trends with the turn toward what he called "second-order cybernetics." In a cluster of papers written by the mid-1970s—"Notes on an Epistemology for Living Things," "On Constructing a Reality," "Cybernetics of Epistemology," and "Objects: Tokens for (Eigen-) Behaviors"—von Foerster catalyzed new thinking about the deeper cognitive implications of "circular causality." Essentially, von Foerster tweaked the engineering discourse of positive and negative feedback toward the recognition of self-reference as an ineluctable form of operation in systems in general, and particularly at the basis of anything worthy of being called cognition, whether the system at hand was natural or technical. The crucial conceptual shift was a movement from first-order cybernetics' attention to homeostasis as a mode of autonomous self-regulation in mechanical and informatic systems to concepts of self-organization—especially as that notion captured the apparent self-ordering and self-regulation of bodies and minds—and to self-reference and autology as the abstract logical counterparts of recursive operations in concrete and worldly systems.

Befitting his adolescent apprenticeship as an amateur magician, Heinz von Foerster's professional papers, many of them reading texts for conference audiences, in addition to the usual scientific diagrams and mathematical equations, are consistently peppered with amusements, puzzles, and paradoxes.[2] Delivered by a mildly manic discursive persona—at times one almost sees the baggy sleeves of the sorcerer's cloak flapping—these papers are intensely focused on entertaining audiences often assumed to be skeptical about their propositions. For scientific papers they are remarkably high rhetorical performances. Serious arguments about matters of biological computation, system/environment interrelation, and perceptual and cognitive construction typically turn on a presentational rhetoric that places them, as it were, on a magician's stage and presto! turns them into visible shapes. Yet the discursive style of this prestidigitating persona is more than a mannerism. It embodies the "natural magic" of cybernetics itself.

With von Foerster's writings, cybernetic discourse begins in earnest the project of taking itself as its own object. With his second-order turn, matters of circular form and operation break out of philosophical and literary treatment (as "reflexivity") and into scientific discussion (as "recursion"). Amazing though it may seem, one discovers that whereas circularity is death (by infinite regress) to *structures*, it is life (by autonomous self-regulation) to *systems*. Von Foerster's work renders paradoxical propositions, recursive forms, and self-referential operations available at once to rational and aesthetic, scientific and literary view. A major and early instance of von Foerster's rhetorical persona discursively embodying cybernetic tenets regarding recursion, epistemological paradox, and self-reference is "On Self-Organizing Systems and Their Environments," delivered in 1959 and first published in 1960. This important paper figures prominently in the work of systems thinkers as disparate as Henri Atlan, Francisco Varela, and Niklas Luhmann and anticipates chaos theory by two decades with its seminal and exhilarating presentation of the "order-from-noise" principle.

Written a decade before the self-referential or neocybernetic turn in his own work, however, this essay still resides in the milieu of first-order cybernetics. It is still large with the heuristic extension of Shannon's information theory as a relay between energy and information; physics and biology; thermodynamical, living, and computational systems. Nevertheless, "On Self-Organizing Systems and Their Environments" prefigures von Foerster's later papers' epistemological turn toward matters of cognitive recursion, in synch with the simultaneous emergence of the concept of biological autopoiesis in the early 1970s, for which discursive event von Foerster was literally the institutional midwife.[3] "On Self-Organizing Systems and Their Environments" is centered on a pre-autopoietic concept of self-organization and the possibility of conceptualizing it through the mathematical theory of communication as the emergence of order from noise.

But to get to this exposition, its audience must first process at some length two conceptual paradoxes, two "fallacies" concerning the definition of self-organizing systems, both of which are subjected to mock deployments of *reductio ad absurdam* arguments. Tying these propaedeutic rhetorical paradoxes together with the essay's subsequent work-up of entropy, information, and redundancy in the interrelations of self-organizing systems and their environments is Maxwell's Demon, a scientific thought-experimental entity custom-made for von Foerster's conceptual whimsy. A preliminary review of the demon's development from its creation in 1867 to its return performance in von Foerster's 1959 paper will allow us to measure the aptness of von Foerster's appropriation of the demon for systems theory, as well as the poetic justice of this collaboration of scientific pranksters. The "magical" or daemonic side of von Foerster's rhetoric—the significance of the paradoxes and other meta-logical performances through which he utters his arguments—can then be more clearly factored into the neocybernetic concepts and methods of his later cognitive papers. Von Foerster's importance for and imprint on the social systems theory of Niklas Luhmann will then emerge with particular clarity.

Protocybernetics in Maxwell's Demon

At a conference entitled "Self-Organizing Systems," von Foerster begins a paper entitled "On Self-Organizing Systems and Their Environments" with the following thesis: "There are no such things as self-organizing systems!"[4] As we noted above, this is the first of the two paradoxes or, more precisely, mock fallacies in this paper. With an obvious allusion to the classical role of Maxwell's Demon as a challenge to the "rule of entropy" in closed physical systems, he asks, ironically, whether "there is not a secret purpose behind this meeting to promote a conspiracy to dispose of the Second Law of Thermodynamics."[5] Maintaining ironical posture, with this remark von Foerster implicitly plays devil's advocate to Maxwell's Demon, as the demon's scientific fame was first established as a successful conceptual antagonist to the "heat death" scenarios spun off from the second law by William Thomson, Hermann von Helmholtz, and many others. Classical thermodynamics declared that physical organization, the orderliness of material/energetic systems, spontaneously deteriorates over time (the rule of entropy) and that its restoration must always be paid for with new contributions of energy. If autonomous self-organization can occur without energic input from or entropic output to the environment, in Maxwell's famous words, this would indeed "pick a hole in the 2nd law." So von Foerster sets up his first paradox by placing the notion of self-organizing systems within a thermodynamic frame, ostensibly like the hermetic enclosure Maxwell used in 1867 to circumscribe the environmental relations of the abstract or hypothetical system into which he first placed his thought-experimental demon.

Maxwell's Demon began as a practical conceit. In this scientific allegory, the daemonic agent personifies the conceptual manipulation of scientific models of physical systems. Viewed historically, the demon began as a scientific fiction but has evolved into a supple theoretical fact with real consequences for the development of modern science. The demon's successful run has rested on its conducting thought experiments in fruitful directions. And few heuristic entities from the annals of scientific modeling have been as thoroughly anthropomorphized—not just brought to conceptual life but also endowed with a narrative career—as Maxwell's Demon. For instance, for several decades around the mid-twentieth century, the demon was declared to be dead—that is, to be a dead scientific metaphor, no longer able to generate plausible challenges to the second law of thermodynamics. Fortunately for the demon, cybernetics revived it for heuristic duty in information theory and computer science, and it can still be sighted today in odd corners of both scientific and popular culture.[6] The demon is clearly more than an academic curiosity or cultural antique: it is the Elvis of Victorian thermodynamics.

On December 11, 1867, Maxwell wrote to fellow physicist Peter Guthrie Tait with a way "to pick a hole" in the second law of thermodynamics, that "if two things are in contact the hotter cannot take heat from the colder without external agency." That is, when physical systems are left to themselves, one always observes heat to move (thermo-dynamics) from hotter to colder bodies. Maxwell invented the demon to reverse, in theory at least, this entropic drift of things. To restore energy to a closed system is potentially to restore it, if the system is then opened, to the environment of that system as well.[7] As first materialized in Maxwell's letter to Tait, the demon takes the form of an *internal* agency within a sealed and partitioned chamber containing a gas: "Now let *A* & *B* be two vessels divided by a diaphragm and let them contain elastic molecules in a state of agitation."

Despite their temperature differential, among the molecules in both chambers "there will be velocities of all magnitudes." It is the random distribution of molecular velocities, together with the partitioning (*CD*) of the total system, that provides the opening for an intelligent agent to fix the molecular lottery. "Now conceive a finite being who knows the paths and velocities of all the molecules by simple inspection but who can do no work, except to open and close a hole in the diaphragm by means of a slide without mass." Supplementing the dynamical system at the microlevel, the demon famously lets only hotter molecules from *B* go into *A*, where it is already hot, and only colder ones from *A* go into *B*, where it is already cold, by which stratagem the second law is violated in that *the hotter vessel takes heat from the colder without external agency.* The demon works on the *inside* of the system but (according to the total conceit of the thought experiment) without adding any energy to the system. Merely by the intelligent sorting of statistical variations in already energetic particles,

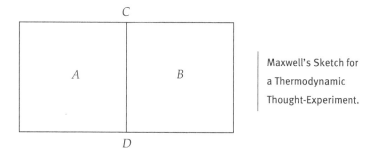

Maxwell's Sketch for a Thermodynamic Thought-Experiment.

the demon reverses the probable drift toward thermal equilibrium and, so the legend goes, saves the world from the specter of entropy.

The demon's career has taken it from the closed systems abstracted from environmental contingencies idealized by classical thermodynamics to the models of self-organizing and autopoietic systems founded on variably open and closed boundaries and structural couplings of systems to each other and to their environments, relationships crucial in cybernetic models of the interfaces among physical, technological, biological, and social systems. As a figure accompanying the postclassical development of the entropy concept, Maxwell's Demon now mediates between the realms of energy and information—classical thermodynamics, quantum physics, and cybernetics. As a broader figure of system functions, as an observer and operator of inner and outer boundaries, the demon can also mediate the distinctions *among* systems, a matter increasingly important to systems theory in a postclassical conceptual milieu where measures of energy (E), thermodynamic entropy (S), and "information entropy" (H) are differentiated and coordinated.

In particular, in its progress from Victorian thermodynamics to contemporary systems theory, one of the demon's signal accomplishments has been to conduct the classical entropy concept from the absolute bondage of thermodynamic closure in material-energetic systems to the contingent liberation of an informatic or statistical virtuality, in which environmental openness to energies and signs is coupled to the operational or organizational closure of self-organizing and autopoietic systems. In Maxwell's original conception, the demon is the internal observer of an isolated physical system, an idealized thermodynamical system with a closed or "adiabatic" outer boundary. By definition, such systems perfectly exclude their environments. In the original case, prior to the arrival of the demon, this exclusion also ensured the eventual thermal equilibrium of vessels A and B, the slurring of their heat differential, and thus the system's submission to the second law. What openness this model possessed was strictly internal: the hole in the partition, which the demon opened or closed to allow selected molecules to pass from one vessel to the other. Yet it was precisely *by partitioning the chamber with a diaphragm and then poking a hole in it,* and so providing the demon with an internal passage point to operate on the basis of its observations, that Maxwell "picked a hole" in the classical idea of thermodynamic entropy in closed systems. In Maxwell's original demon scenario, that is, the external closure of the physical system is countered by the demon's capacity to open and shut the membrane between the vessels. Thus the demon's operations already anticipated the "open closure" of complex systems that go beyond thermodynamical enclosure and use their boundaries to regulate commerce (input/output ratios or perturbation/compensation relations) with their environments.

"On Self-Organizing Systems and Their Environments"

Mock Fallacy 1: There Is Such a Thing as a Self-Organizing System

With his allusion to the second law at the beginning of "On Self-Organizing Systems and Their Environments," von Foerster sets up his first paradox—"I shall now prove the non-existence of self-organizing systems by *reductio ad absurdam* of the assumption that there is such a thing as a self-organizing system"—by placing the notion of self-organizing systems within the frame of classical thermodynamic constraints. These include the "adiabatic shell" (see below), across which no energy may pass. The hermetic closure of this universe from any other universe underscores the importance of *environmental closure* in the classical thermodynamic milieu. The universe of classical thermodynamics is closed not only at the (ideally) sealed outer boundaries of thermodynamical systems, such as heat engines, but also insofar as that universe itself, the environment of all environments, is envisioned as a closed system. Von Foerster framed his model of a self-organizing system with these classical thermodynamic allusions to closed partitionings of *energy*, it seems, precisely to elicit certain systemic *alternatives* to this form of closure. That is, his essay unfolds the repercussions for self-organization for systems that process *information* as well.

Assume a finite universe, U_0, as small or as large as you wish, which is enclosed in an adiabatic shell which separates this finite universe from any "meta-universe" in which it may be immersed. Assume, furthermore, that in this universe, U_0, there is a closed surface which divides this universe into

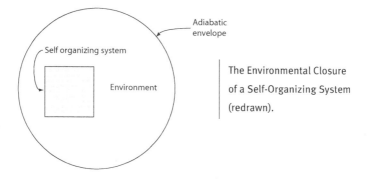

The Environmental Closure of a Self-Organizing System (redrawn).

two mutually exclusive parts: the one part is completely occupied with a self-organizing system S_0, while the other part we may call the environment E_0 of this self-organizing system: S_0 & $E_0 = U_0$. . . .

Undoubtedly, if this self-organizing system is permitted to do its job of organizing itself for a little while, its entropy must have decreased during this time . . . otherwise we would not call it a self-organizing system, but just a mechanical . . . or a thermodynamical . . . system. In order to accomplish this, the entropy in the remaining part of our finite universe, i.e. the entropy in the environment must have increased . . . otherwise the Second Law of Thermodynamics is violated. . . . Hence the state of the universe will be more disorganized than before . . . , in other words the activity of the system was a disorganizing one, and we may justly call such a system a "disorganizing system."[8]

In this passage von Foerster plays fast and loose with, or rather mocks, (1) the noncoincidence of the system S_0's *self*-organization with its "disorganizing" of its *other*, the environment E_0, and (2) the distinction between energy and information. That is, for the moment he equivocates between, on the one hand, thermodynamic entropy as the measure of some real reduction in the sum of usable energy within a material system and, on the other hand, entropy as redefined in Shannon's information theory as a measure of the (dis)order or formal (dis)organization of any system, but particularly of message structures (information) transmitted within communication systems. But the equivocations embedded in this supposed proof bring out the paradoxical punch line of von Foerster's mock fallacy: distinctions are also inclusive of what they exclude. From the holistic perspective that stays attentive to both system *and* environment, the universe U_0 is comprised by the *mutual inclusion* of these "two mutually exclusive parts": "In spite of this suggested proof of the non-existence of self-organizing systems, I propose to continue the use of the term 'self-organizing system,' whilst being aware of the fact that this term becomes meaningless, unless the system is in close contact with an environment, *which possesses available energy and order*, and with which our system is in a state of perpetual interaction, such that it somehow manages to 'live' on the expenses of this environment."[9]

Von Foerster administers these bracing doses of classical thermodynamics and first-order cybernetics, then, to put his audience's material-energetic feet on the ground. To balance the bias toward information and the valorization of order taken as forms *internal* to self-organizing systems, he insists on maintaining conceptual hold on the environment, what Cary Wolfe calls "the pragmatics of the 'outside.' "[10] "Energy and order" are characteristics particularly plain in living systems and in the environments that sustain them, and the metaphorics of this latter passage are more biological than physical. Before the emergence of far-from-equilibrium thermodynamics and dynamical systems theory, considerations of self-organization were typically centered on *living* systems and then extended, not so much to physical or chemical systems as to the biotic and metabiotic natural systems—nervous, psychic, and social—that ramify from the evolution of cells and organisms within material-energetic, thermodynamical environments. Von Foerster's serious point remains indispensable: No system of any stripe can be adequately treated in the absence of the environment it constitutes for itself by emerging as a system. The inseparability of the system/environment dyad is a primary and pivotal premise of cybernetic thought, separating cybernetics per se from the late-classical paradigms of Victorian thermodynamics. To characterize the argument so far, "There is such a thing as a self-organizing system" is a *mock fallacy*—strictly speaking, a paradox. Although certain systems *do* self-organize, or decrease their internal entropy, they do so only in the presence of conditions provided for elsewhere, by environments that lend a necessary other to the self of self-organization.

Mock Fallacy 2: This World Is Only in Our Imagination

The inseparability of the system/environment dyad is also the point of von Foerster's second gambit at the beginning of "On Self-Organizing Systems and Their Environments." But this time he approaches it from a philosophical rather than biophysical vantage. From the side of the "subject" rather than that of the "object," he arranges a kind of Cartesian litmus test for the reality of the environment.[11] For in its first moment, all the Cogito can cognize is its *own* existence, precisely as a *system* capable of *self*-observation: "Perhaps one of the oldest philosophical problems with which mankind has had to live . . .

arises when we, men, consider ourselves to be self-organizing systems. We may insist that introspection does not permit us to decide whether the world as we see it is 'real,' or just a phantasmagory, a dream, an illusion of our fancy."[12] What if the only reality is in fact the self in terms of which the mind carries out its self-organization? If that were the case, "my original thesis asserting the nonsensicality of the conception of an isolated self-organizing system would pitiably collapse. I shall now proceed to show the reality of the world as we see it, by *reductio ad absurdum* of the thesis: this world is only in our imagination and the only reality is the imagining 'I.' "[13] Or again (says von Foerster, honorary cousin of Ludwig Wittgenstein) I shall now make a superannuated philosophical conundrum—disappear![14]

With this buildup, von Foerster musters up one of the more famous icons in the visual rhetoric of cybernetic discourse. In order to problematize the solipsistic notion of the *mind* as "an isolated self-organizing system," von Foerster diagrams a situation of self-referential selves within selves—the Man with the Bowler Hat (MBH). The MBH imagines that he is a self-organizing system *without an environment*:

> Assume for the moment that I am the successful business man with the bowler hat . . ., and I insist that I am the sole reality, while everything else appears only in my imagination. I cannot deny that in my imagination there will appear people, scientists, other successful businessmen, etc., as for instance in this conference. Since I find these apparitions in many respects similar to myself, I have to grant them the privilege that they themselves may insist that they are the sole reality and everything else is only a concoction of their imagination. On the other hand, they cannot deny that their fantasies will be populated by people—and one of them may be I, with bowler hat and everything!
>
> With this we have closed the circle of our contradiction: If I assume that I am the sole reality, it turns out that I am the imagination of somebody else, who in turn assumes that *he* is the sole reality. Of course, [since either of these propositions is "absurd"] this paradox is easily resolved, by postulating the reality of the world in which we happily thrive.[15]

The X-ray image of the Man with the Bowler Hat ostensibly depicts the classical solipsist—one who asserts that whatever they perceive (here, those two gentlemen over there) exists only inside their own mind and that, therefore, *they* are the only thing in existence. Even if idealist philosophies of the transcendental subject do not rule such notions out, however, according to the primary premise of systems theory rehearsed in the first paradox, in the world of

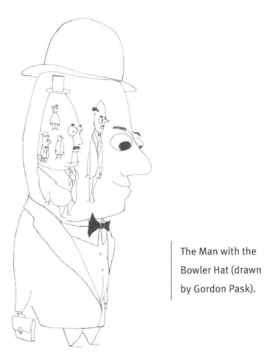

The Man with the
Bowler Hat (drawn
by Gordon Pask).

real systems the nonexistence of an environment is simply not forthcoming. So solipsism is not the real problem but rather a metaphor for a real propensity to concentrate on the system concept without maintaining sufficient consideration of its necessary counterpart, its environment. Or again—to be adequate to its object, systems thought has to proceed from the full situation constituted by the system and its environment and the dynamic maintenance of the boundary that distinguishes them.

In the 1959 paper, then, von Foerster uses the MBH to reaffirm the proposition just established, to underscore again, from the side of the subject, the real existence of the environment over against a too narrow interest in the system concept per se. Von Foerster deconstructs the paradox of solipsistic self-descriptions by reframing it within a world created by the mutual observations of two or more (mock) solipsists. To state the moral of this parable in second-order systems parlance: if autopoietic systems can *not* be environmentally closed systems (as thermodynamic systems ideally can) and if the mind is an autopoietic system (the psychic system), then solipsism—"pure" self-reference—has no way to even begin to operate. Solipsism is thus dismissed as cognitive repression of the environment.

Still, the paradox is so easily resolved that we almost fail to notice that what we have just been presented is another bit of conceptual sleight of hand, a mock proof, a rhetorical argument masquerading as a logical *reductio*. As Katherine Hayles has noted, although without, I think, appreciating the full significance of her observation, "Von Foerster himself seemed to recognize that the argument was the philosophical equivalent to pulling a rabbit from a hat."[16] But the worldly environment within which all cognitive systems are embedded was not really in serious doubt, only the rigor with which it got itself factored into systems thinking. What von Foerster implies with his magical fallacy is that we still have a hard time taking for real that all *knowledge* of the environment depends upon the specific realities of the systems that observe it. The *systemic* reality of the environment is to be both the precondition *and* the product of an observing system. This is the self-reference bound up in any presentation of something beyond the self. The MBH conveys an ironic message to those who would affirm that when they observe the world and state their truths about it, they are *not* a part of the reality they are describing, that they are not embedded within their own descriptions. *This* denial would be the sort of reverse solipsism of the epistemological positivist that posits a contextless context of knowledge—as if the world could be known without the existence of local and embodied knowers to carry out the knowing. So the "subjective" self/other difference of the second paradox doubles the "objective" system/environment coupling developed in the first paradox, allowing for the different, environmental reality of the other to arise a second time, this time out of the self-reference of the self as observing system.

In the 1959 paper, von Foerster concludes the bowler hat interlude with a methodological proposition in the form of a discursive aside: "Having reestablished reality, it may be interesting to note that reality appears as a consistent reference frame for at least two observers."[17] And if we examine the MBH carefully, we see that he constructs in imagination *two* observers, one of which confirms his reality by imagining *him*. The total figure, then, prefigures what second-order cybernetics will call the observation of observation. Through a reticulation of levels of observation, it presents a second-order view of the epistemological situation, which reveals that the solipsist, in turn, exists both as the figment of its own figments and as the observation of another observer.

For the solipsist to confess that its own existence depends on the reality of an other in its environment is a parable of the overcoming of ontological idealism by epistemological constructivism. Taken as a diagram of recursive cocognition, then, the MBH is not a sterile *mise-en-abyme*, hall of mirrors, or vicious circle of the same all over again. Literally considered, the MBH does not simply self-reiterate in a mode of infinite regress; rather, an other is interposed between the self and its reflection. This is better understood as a productive oscillation, as the image of a mutual embedding of the other within the self and the self within the other. The message is that the reality one can know depends on the communication of reality from one observer to another, which depends on "a consistent reference frame" within which "at least two observers" are embedded so as to construct a conversation about that reality.

Two Maxwell's Demons and the "Order-from-Noise" Principle

As we have just seen, "On Self-Organizing Systems and Their Environments" adumbrates the explicit epistemological constructivism of von Foerster's later essays with a systems theory built on the premises of recursively structured system/environment couples. A multiplicity of mutually reinforcing observers maintains relationships and states of stable cross-systemic resonance von Foerster will come to call "eigenbehaviors," at both the biological and the social level. Thus it is fitting that when Maxwell's Demon makes a literal entrance later in the paper, it does its self-organizing not as a solo act within a closed system, but instead as a duo of demons collaborating to self-organize within a system/environment dyad under their mutual observation.

This time, the matter of entropy is treated cybernetically, in its information-theoretic redescription as a measure of the relative order or disorder of an information source: "Order has a relative connotation, rather than an absolute one, namely, with respect to the maximum disorder the elements of the set may be able to display." Von Foerster advances Claude Shannon's definition of redundancy (R) as a measure of the order of a system, derived by subtracting from unity or maximum order the relative entropy, "the ratio of the entropy H of an information source to the maximum value, H_m, it could have while still restricted to the same symbols."[18]

Von Foerster works through further mathematical models that would obtain for a system whose self-organization is observed in these terms. The self-organizing operations of the system, it turns out, can be viewed from either side of the ratio. For instance, if the maximum entropy H_m is held constant, then the system will be self-organizing if the entropy H can be decreased. This represents the "internal" view of the situation, and von Foerster slips the first demon into the demonstration to underscore its boundedness on the inside of the system: "Since all these changes take place internally I'm going to make an 'internal demon' responsible for . . . shifting conditional probabilities by establishing ties between elements such that H is going to decrease."[19] However, if instead of the maximum entropy H_m, the given entropy of the system H is held constant, "we obtain the peculiar result that . . . we may have a self-organizing system

$$R = 1 - \frac{H}{H_m}$$

Claude Shannon's Definition of Redundancy.

before us, if its possible maximum disorder is increasing."[20] In other words, it is also possible to increase the order of a system *by maintaining its entropy while increasing its complexity*: "Clearly, this task of increasing H_m by keeping H constant asks for superhuman skills and thus we may employ another demon whom I shall call the 'external demon,' and whose business it is to admit to the system only those elements, the state of which complies with the conditions of, at least, constant internal entropy."[21]

However, in both of these limit cases one demon is chained while the other does all the work. The more common situation will be one "where both demons are free to move"[22] or—translating this mathematical allegory of self-organization into somewhat plainer terms—one where systematic processes and environmental resources are structurally coupled to each other in a way that maintains the system and under favorable conditions enables it to reduce its entropy and/or increase its complexity. The paradoxical boundary where system and environment meet is marked by the "greater than" sign indicating the higher order of the system relative to the lesser order but greater complexity of its environment. In this parable of productive mutuality, what I also like to read as a fable of academic interdisciplinarity, "if the two demons are permitted to work together, they will have a disproportionately easier life compared to when they were forced to work alone."[23]

Now von Foerster makes a famous final move that synthesizes the energic and informatic potentials of his system-theoretical demons and derives the *order-from-noise* principle from "the double linkage between the internal and external demon which makes them entropically (H) and energetically (E) interdependent." It is not just a matter of "negentropy," as that term was used at the time to identify informational order with negative entropy. Self-organizing systems can also translate the *external noise* of their environments into systemic gain: "Thus, in my restaurant self-organizing systems do not only feed upon order, they also find noise on the menu."[24] With the formulations that "reality appears as a consistent reference frame for at least two observers" and "a self-organizing system feeds upon noise," von Foerster both shapes and integrates the future courses of second-order and self-organizing systems theory. Citing Ilya Prigogine's work on dissipative structures in the discussion that followed his 1959 presentation,[25] he inspires the theoretical bioinformatics of Henri Atlan, taken up by Michel Serres, and anticipates broad future developments in chaos

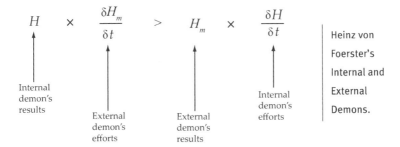

theory, the actor-network theory of Bruno Latour, and the autopoietic systems theories of Maturana and Varela and Niklas Luhmann.[26]

To demonstrate the order-from-noise principle, von Foerster plays a sophisticated game of stacking blocks. Conjured up to probe Shannon's mathematics of information in light of a dual system-and-environment approach to the self-organization concept, the exposition of the order-from-noise principle is done up with magnetic cubes reminiscent of Charles Howard Hinton's tesseracts or hypercubes.[27] The blocks are prepared by gluing to their faces magnetized squares with either the north or south pole of the magnetization facing outward. Out of the ten possible arrangements, various cubes will have more or less propensity to couple as their surfaces present opposite or the same magnetic poles. The probability is high that attracted surfaces will indeed fasten to each other. If a large number of cubes other than those that offer all north or all south poles, so only repel each other, is placed in a box in a jumble and given a vigorous shake, the cubes will assume a more ordered formation. "The entropy of the system has gone down, hence we have more order after the shaking than before." Indeed, if one works with a population of cubes with a particularly favorable arrangement of facet polarities, von Foerster proclaims, "you may not believe your eyes, but an incredibly ordered structure will emerge, which, I fancy, may pass the grade to be displayed in an exhibition of surrealistic art."[28]

How could this happen? "The shaking, of course—and some little demons in the box." Von Foerster's daemonic duo outdo Maxwell's Demon, as, flashlight or no, it had to sort by detecting particles, calculating their trajectories, selecting and rejecting candidates for transfer, then opening and closing the "slide without mass" over the aperture in the diaphragm. In von Foerster's self-organization scenario, the environment donates its energies to the processes increasing the system's order. As with the minds of children, for instance, ordering can emerge when random energies are distributed among relatable elements such as building blocks and given enough time to play. Just as no additional

Jumbled Boxes
(drawn by
Gordon Pask).

energy was imported into Maxwell's Demon's system, "No order was fed to the system [of the blocks], just cheap undirected energy; however, thanks to the little demons in the box, in the long run only those components of the noise were selected which contributed to the increase of order in the system."[29]

Best of all, in this scenario these offspring of Maxwell's Demon are not entirely imaginary: "The co-operation of our demons" is guaranteed because they are not interlopers from elsewhere, but are themselves "created along with the elements of our system, being manifest in some of the intrinsic structural properties of these elements."[30] Von Foerster's self-organizing demons are as allegorical as Maxwell's, but they are not merely thought-experimental. Rather, they are pragmatic personifications of the structural attributes of material-energetic elements, the building blocks of systems and their theories. Whereas Maxwell's Demon was first envisioned as an idealized supplement to the description of a closed entropic system, von Foerster's demons derive from cybernetic models of the radical distinction yet operational coupling between energy and information in complex systems. Thus, with von Foerster, the demon may be seen to have entered a further, post-computational phase of his career, now as a messenger of deterministic chaos and as a mediator between information theory and cybernetic systems theory.

Self-Organization
from Noise (drawn
by Gordon Pask).

Neocybernetics of Social Systems Theory

We may take it that the world undoubtedly is itself (i.e. is indistinct from itself), but, in any attempt to see itself as an object, it must, equally undoubtedly, act so as to make itself distinct from, and therefore false to, itself. In this condition it will always partially elude itself.—George Spencer-Brown, *Laws of Form*

From one paper to the next in *Understanding Understanding*, a certain reshuffling of the same deck occurs as need arises. But von Foerster's discursive recursions are not merely repetitive or perfunctory: his discourse evolves, and recurrent elements take on new roles and bear new meanings. Take perhaps the most accessible and broadly disseminated rendering of von Foerster's insights into recursive neural computation and what Maturana and Varela would soon call the "organizational closure" of autopoietic systems, the 1973 paper "On Constructing a Reality." In keeping with the rhetorical pattern I described above, it begins with a humorous and erudite literary allusion, then segues to a series of perceptual puzzles eliciting "blind spots" in the sensorium before settling into its central arguments regarding neural computation and the "double closure" of cognitive systems.

"On Constructing a Reality" is a seminal annunciation of second-order cybernetics, precisely as a constructivist theory of cognition. As one now says in the vocabulary of George Spencer-Brown, "On Constructing a Reality" *re-enters* the form of cybernetic observation into its own form. Von Foerster later coins the slogan "the observation of observation" for this mode of cybernetic self-reference, and "On Constructing a Reality" prefigures this slogan with its logical derivation of cognition as recursive computation.

Computation is generalized to mean any process or algorithm that transforms or recodes stimuli or data presented to it: "Computing (from *com-putare*) literally means to reflect, to contemplate (*putare*) things in concert (*com*), without any explicit reference to numerical properties. Indeed, I shall use this term in this most general sense to indicate any operation (not necessarily numerical) that transforms, modifies, rearranges, orders, and so on, observed physical entities ('objects') or their representations ('symbols')."[31] Although the term "autopoiesis" does not appear in it, several of Humberto Maturana's works, preliminary to that coinage co-authored with Varela, are cited in it, and the form of the concept of autopoiesis—self-referential recursion bounded by operational closure—is limned throughout the essay.

Using recursion as a skeleton key to unlock a range of complex self-referential systems, von Foerster's second-order cybernetics arrived at a general discourse

Cognition as Recursive Computation.

of operational circularity by turning cybernetic thinking upon itself. Luhmann has written about cybernetics in general that "the first innovation was the re-discovery of the circle as, at the same moment, a natural and technical form."[32] When viewed in this wider context, much of Luhmann's social systems theory extends directly from von Foerster's contributions to the recuperation of self-reference.[33] In "Notes on an Epistemology for Living Things" von Foerster sketches the ways that, for most of a century, hard scientific thought has been forced to acknowledge the paradoxes of observation.[34] Forcing the epistemologi-cal issue of second-order cybernetics, Luhmann puts the paradox point blank in "The Cognitive Program of Constructivism": "It is only non-knowing systems that can know; or, one can only see because one cannot see."[35] However, how *can* a discourse of knowledge founded on a concept of self-reference, which implies the operational closure of subjects of knowledge reconceptualized as "observing systems" cut out from their surrounding environments, result in anything but a short circuit or an infinite regress?

To shift epistemology to an explicitly recursive system/environment para-digm forces a cascade of repercussions. This cognitive regime bars any tradi-tional form of empirical or realist representationalism, any simplistic notion of knowledge as a mechanics of linear inputs and outputs. Redescribed as the production of an observing system, cognition is rendered as a contingent op-erational effect rather than assumed as a free-floating or even disembodied agency. The boundary between "subject" and "object" is re-cognized as both an ongoing product of and an impassable limit to the operation of the system. A system boundary never just is, ontologically, but is always coming into being as part and parcel of the system's total autopoiesis. As a self-referential product of the system's operational enclosure, the boundary guarantees that the system is autonomous, or "information-tight." The environment as "object" cannot enter into the system in the mode of its own being, cannot dictate to the system. What it can do is perturb its observer in such a way that the system reorganizes its own elements to compensate, which compensation must then count for the system's cognition of the object. One of the more scandalous ways to express this situation appears as Proposition 11 in von Foerster's "Notes on an Epistemology for Living Things": "*The environment contains no information; the environment is as it is.*"[36] That is to say, it is only self-referential observing systems that can construct environments in the mode of information; the construction of these

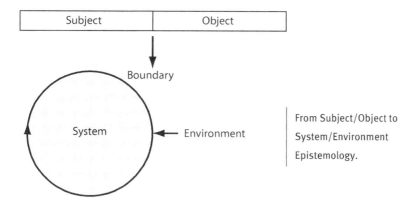

From Subject/Object to System/Environment Epistemology.

descriptions is solely the system's affair. "*The environment is as it is.*" However, there is always more than one observing system; or again, cognition is always also a social affair. Observing systems in communication may be observed to arrive at *eigenbehaviors*—that is, mutually to stabilize and reinforce perceptions autonomously achieved.[37]

The virtual boundaries of social and psychic systems are produced and reproduced by the forms of distinction those same systems construct in the medium of meaning—say, between self and other, between inclusion and exclusion—that render those systems operable at any given moment, maintainable from one moment to the next, and sufficiently distinct from their environment and other systems to maintain operations. Luhmann writes: "Boundaries can be differentiated as specific mechanisms with the specific purpose of separating yet connecting. They assume this function via particular performances of selection."[38] Psychic and social identities coalesce around a system's probable reiteration of the same selections from a given repertoire of possible distinctions and may be transformed when different selections ramify into a new norm or new options enter the repertoire of possible distinctions. But because the inevitable effect of a system's history of self-bounding through cognitive selections is to have excluded, at least for the time being, other forms of possibility, Luhmann goes on to note that "a contact mediated by boundaries cannot convey to any system the full complexity of another, even if its capacity for processing information would otherwise be sufficient."[39]

The figure on page 54 diagrams the operation of reentry as a model of recursive cognition in the second-order cybernetic description of observing systems. This is how one constructs a reality: an observing system **S**, a necessarily self-referential form, creates epistemological space for itself by reentering the virtual form of its own bounded distinction from the environment **B/E** into itself as

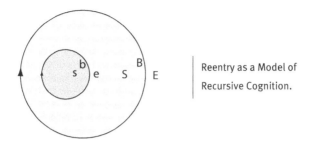

Reentry as a Model of Recursive Cognition.

the virtual border **b/e**, which it can then use to make distinctions between self and other. It can then, at any given moment, construct its selective knowledge **e** as a reduction of the complexity of the environment **E** rendered through its own repertoire of distinctions. We see that our knowledge of **E** will always be a somewhat lesser version, **e**. But that's still saying something, and **S** can also proceed to test its knowledge, its internal model **e,** against other versions constructed at other moments. In this way we see that **e** is not a static production but an ongoing, recursively refreshed computation. And recursive processes, like rolling hoops or gyroscopes, are self-stabilizing—they tend to find their own balance.

In "On Constructing a Reality," focused on a comparable model of "double closure," von Foerster draws from his discussion of neural computation a conclusion that we can also apply to the model in figure 10. According to "the postulate of cognitive homeostasis," our self-referential constructions of the world are rendered relatively stable (and not merely arbitrary) because as a result of its own recursive self-corrections, "The nervous system is organized (or organizes itself) so that it computes a stable reality."[40] But still, how does this self-referential construction of epistemological constructivism differ from traditional notions of idealism, in which the world of objects was also presented as a product of the mind's own activities? As Luhmann observed at the outset of "The Cognitive Program of Constructivism," "It is only in our century that the name 'idealism' has been replaced by 'constructivism.' . . . Insofar as constructivism maintains nothing more than the inapproachability of the external world 'in itself' and the closure of knowing—without yielding, at any rate, to the old skeptical or 'solipsistic' doubt that an external world exists at all—there is nothing new to be found in it."[41] To earn its epistemological spurs as a true and pivotal redescription of our knowledge of knowledge, that is, neocybernetic constructivism must demonstrate its actual operationality, its *social* productivity beyond any singular mind's phenomenality.

This is one reason why von Foerster concludes "On Constructing a Reality" with a coda acknowledging that his foregoing discussion could plausibly be dismissed as a plea for solipsism, if it were to be delimited by old philosophical habits, "the view that this world is only in my imagination and the only reality is the imagining 'I.' Indeed, that was precisely what I was saying before, but I was talking only about a single organism. The situation is quite different when there are two."[42] We see that the fallacy of solipsism always was an aberration but not a paradox: it was an inference logically induced by the idealization of singularity, the residual monotheologism that allowed the conception of disembodied observations—that is, the conception of the possibility of systems without environments, and thus the possibility of a system unaccompanied by other systems. The real paradox is that we could have gone so long imagining that there could be not just minds without bodies or worlds, but also minds in the absence of other minds.

As we recall, in "On Self-Organizing Systems and Their Environments," the Man with the Bowler Hat prefigured but did not unfold the matter of cognitive self-reproduction, or the autopoiesis of the psychic system. The main thrust of that essay was self-organization as the emergence of order from noise in material-energetic systems. When the MBH makes an encore fourteen years later, it is no longer put to work to support the reality of the environment per se, but rather to support a view of that reality that now rests solely and explicitly on its cognitive co-construction by multiple observing systems. Misreadings of the MBH are more likely to occur in the context of this paper, however, because here its epistemological exposition is relatively compressed. Having just spent an entire paper formulating the recursive nature and operational closure of the nervous system, in a way now directly (if still implicitly) tied to the propositions of second-order cybernetics, von Foerster has less interest in using the MBH to satirize and deconstruct the architecture of solipsism. Rather, "solipsism" is morphed into "irresponsibility" and blasted into a systems dimension that provides for its self-overcoming. Whereas in the 1959 paper the mantra of the solipsist had been, "I insist that I am the sole reality,"[43] the entire 1973 paper revolves around an initial constructivist postulate—"The Environment as We Perceive It Is Our Invention"[44]—that would seem considerably to up the solipsist's ante.

Instead, from here we can see more clearly why the second paradox in "On Self-Organizing Systems and Their Environments" was a *mock* fallacy, a statement that only seems to oscillate between true and false. Rather, it *is* true, just not true enough. Once again, as in the first paradox regarding systems without environments, *it captures something true but not sufficient about reality.* Namely, *what is true* (that is, as true as possible under constructivist constraints on the concept of "truth") is that we do have to construct reality—"the environment as we perceive it *is* our invention." But *what is insufficient* is the implication that when we do so, we are in the solipsistic situation of going it alone. In "On Constructing a Reality," the grammatical doubling of the constructivist postulate ("the environment as *we* perceive it") carries the epistemological weight. Whereas solipsism proceeds in the singular, constructivism proceeds in the plural. Solipsism is transcended not by negating its proposition but by forcing the complexity of the multiple out of its unitary simplicity. Epistemology proceeds from classical capture by the singularity of the knowing mind to multiple knowledges in social contexts.

Second-order cybernetics sees instead a world so constructed that any single observer's observations may be rendered stable from moment to moment by the structural couplings and recursive conversations of *its* multiple observers. Just as all nervous systems and all organisms that possess them within themselves are virtual consortiums of multiple autopoietic systems, so are all observers bound into (what Varela calls) "observer-communities" within which (what Luhmann calls) *social* autopoiesis—the ongoing self-production of and self-maintenance of communication—produces (what von Foerster calls) *eigenvalues*—that is, stable yet mobile and multiple recursive consensuses about shared environments.

By way of conclusion, let us read some of Luhmann's opening moves in *Social Systems* to appreciate the force with which von Foerster's text inflects Luhmann's systems theory, as well as the extent to which Luhmann's theory represents one of the most important ways that von Foerster's pioneering work has come to further fruition. Luhmann's "Introduction: Paradigm Change in Systems Theory" states the following:

> The theory of self-referential systems maintains that systems can differentiate only by self-reference, which is to say, only insofar as systems refer to themselves (be this to elements of the same system, to operations of the same system, or to the unity of the same system) in constituting their elements and their elemental operations. To make this possible, systems must create and employ a description of themselves; they must at least be able to use the difference between system and environment within themselves, for orientation and as a principle for creating information. Therefore self-referential closure is possible only in an environment, only under ecological conditions.[45]

The footnote here is to von Foerster's "On Self-Organizing Systems and Their Environments," one of the essential points of which was that in the vogue for notions of self-organization then abroad at the turn of the 1960s, systems theory also had to discipline itself against traditional biases toward the singular *self* of self-organization at the expense of its multiple and heterogeneous other, its environment. Luhmann's phrase "ecological conditions," then, is neatly parsed to mean "environmental contingencies."

Luhmann continues: "The environment is a necessary correlate of self-referential operations because these out of all operations cannot operate under the premise of solipsism."[46] The footnote here is to von Foerster's "On Constructing a Reality," implicitly to the coda we were just discussing, where the treatment of solipsism is itself a compressed version of a longer passage from the earlier "On Self-Organizing Systems." This compounds the extent to which Luhmann is leaning on von Foerster's provision of systems-theoretical premises that predate the explicit turn toward second-order recursions. But Luhmann's passage goes right on to reinvoke matters of distinction and reentry that we can appreciate as the Spencer-Brownian side of von Foerster's legacy: ". . . cannot operate under the premise of solipsism (one could even say because everything that is seen playing a role in the environment must be introduced by means of distinction). The (subsequently classical) distinction between 'closed' and 'open' systems is replaced by the question of *how self-referential closure can create openness*."[47] This conceptual framework of a paradoxical yet recursively operational *open closure* spanning the autopoietic spectrum from biotic to metabiotic instances is the crux of the radical neocybernetic reorientation of systems thinking around a paradoxical "sublation" of the classical polarity between "closed" and "open" systems, over which thinkers as astute as Hayles continue, for whatever reasons, to stumble. Polarities aligning humanism with openness and antihumanism with closure shift to strategies for an adequate conceptual grasp of the complex *posthumanism* of neocybernetic thought.[48]

In the immediate continuation of this passage, Luhmann offers his posthumanist formulation of the epistemological boundaries or observational constraints Hansen in this volume foregrounds in Varela's "ethico-ontological" precept respecting the autonomy of autopoietic observing systems: "Here too one comes to a 'sublation' of the older basic difference [between system and environment] into a more complex theory, which now enables one to speak about the introduction of self-descriptions, self-observations, and self-simplifications within systems. One can now distinguish the system/environment difference as seen from the perspective of an observer (e.g., that of a scientist) from the system/environment difference as it is used within the system it-self, the observer, in turn, being conceivable himself only as a self-referential system."[49]

In "On Constructing a Reality" von Foerster presents his own "ethico-ontological" understanding of neocybernetic co-constructivism by forcing the social issue out of the biological and neurological factors of operational closure, wringing from the MBH the cognitive confession that "reality = community." He unfolds from the pair of internal observers in the MBH a pair of co-constructing social imperatives—two performative utterances bound together by a feedback loop. The "*ethical imperative*" is "Act always so as to increase the number of choices"; the "*aesthetical imperative*" is "If you desire to see, learn how to act."[50] If ethics concerns the *operations* systems select for themselves, especially insofar as these systems may be considered as self-referential or "reflective" entities, aesthetics concerns the role of *observations* in guiding those operations. To shift into the conceptual mode of second-order systems theory, then, is to reform classical dyads such as part/whole divisions or subject/object dichotomies, in order to reobserve and grasp the complexly supplementary rather than merely exclusionary relations of system/environment couples and to re-cognize the borders that self-organizing, now *autopoietic,* systems must use in order to operate.

Simply put, taking systemic self-reference seriously subverts the notion of "objective" material or philosophical foundations. And those who believe in the actual or possible possession of "objective truth" can only see in such a deconstruction an occasion for its flip side, that other mirage called "subjec-tive relativism." This negative oscillation of universal truth and individual falsity is the infinite regress one enters by *avoiding* the productive encounter with paradox. On the plus side, observers can now choose to construct them-selves as inside and a part of, as well as outside and detached from, the systems and environments they observe. And if we were to convince ourselves that the whole show—the virtual integration of all biotic and metabiotic systems into a global environment stretching out to the boundless cosmos—is at all times *collectively* self-bootstrapping—that is, held up by the self-maintenance and complex interlocking of all of its strange loops—we might start to be-have in less merely selfish ways. Who knows, we might even begin to over-come the larger *social* solipsisms that are currently wreaking destruction on social systems and their natural environments. While Luhmann himself re-mained skeptical of such a recuperative reading of systems theory, von Foerster did not.

Notes

1. Firsthand accounts of these events are given in von Foerster with Poerksen, *Understanding Systems*, 135–40, and Brand, "For God's Sake, Margaret."

2. See his family reminiscence, "Introduction to Natural Magic." An illustrated biography of von Foerster (in German) is Grössung, Hartman, Korn, and Müller, *Heinz von Foerster 90*; p. 16 shows a snapshot of von Foerster and his cousin as teenaged magicians.

3. In "The Early Days of Autopoiesis" in this volume, Varela narrates the history of the first appearance of "autopoiesis" in a 1973 Chilean publication by Maturana and Varela, *De máquinas y seres vivos*, and von Foerster's vetting of its first appearance in English; Varela, Maturana, and Uribe, "Autopoiesis." Von Foerster then publishes as Biological Computer Laboratory Report 9.4 (September 1, 1975) their extended treatment, *Autopoietic Systems*, which eventually appears as *Autopoiesis and Cognition*. See also Zeleny, "Autopoiesis."

4. Von Foerster, "On Self-Organizing Systems," 1. The article was first published in Yovits and Cameron, *On Self-Organizing Systems*, 31–50.

5. Von Foerster, "On Self-Organizing Systems," 1.

6. In the standard anthropomorphic idiom, the editors of a compendium of scientific papers on Maxwell's Demon write: "The life of Maxwell's demon can be viewed usefully in terms of three major phases" (Leff and Rex, *Maxwell's Demon*, 2). The Beatles' "Maxwell's Silver Hammer" is a likely atavism; cf. "Maxwell's Demon" as Brian Slade's (Jonathan Rhys-Meyers) glam persona in *Velvet Goldmine*.

7. Because of the notion of using a demon to *generate* energy out of thin air rather than, with Maxwell, to restore extant energy (by the first law) from unusable to usable form, the demon is sometimes simplistically presented as the operator of a perpetual motion machine. For more on Maxwell and his demon from a literature and science perspective, see Clarke, *Energy Forms*, ch. 4; and Hayles, "Self-Reflexive Metaphors." For more on the culture of thermodynamics, see Prigogine and Stengers, *Order Out of Chaos*; Stengers, *Power and Invention*, esp. "Turtles All the Way Down" (60–74); and Clarke and Henderson, *From Energy to Information*, part 1. A new theory of thermodynamics in its cosmic and biospheric contexts is given in Schneider and Sagan, *Into the Cool*.

8. Von Foerster, "On Self-Organizing Systems," 1–2; my elisions of the mathematical notation in the original.

9. Ibid., 3.

10. Wolfe, *Critical Environments*, subtitle.

11. "You remember René Descartes. . . . 'Am I, or am I not?' He answered this rhetorical question with the solipsistic monologue, 'Je pense, donc je suis'" (von Foerster, "Ethics and Second-Order Cybernetics," 297).

12. Von Foerster, "On Self-Organizing Systems," 3–4.

13. Ibid., 4.

14. See von Foerster with Poerksen, *Understanding Systems*, where von Foerster states about Ludwig Wittgenstein, "He was an honorary uncle, a *Nennonkel*, as they say in Vienna, not a relative, but a very good friend. My mother and his sister, Margarethe, were good friends" (122).

15. Von Foerster, "On Self-Organizing Systems," 4.

16. Cf. Hayles, *How We Became Posthuman*, 133. I am suggesting that Hayles misses the point when she comments, "Although charmingly posed, the argument is logically nonsensical" (133).

17. Von Foerster, "On Self-Organizing Systems," 4.

18. Ibid., 7.

19. Ibid., 8.

20. Ibid., 9.

21. Ibid.

22. Ibid., 10.

23. Ibid.

24. Ibid., 11.

25. Von Foerster, *Understanding Understanding*, 19.

26. Atlan treats von Foerster's order-from-noise principle in "Hierarchical Self-Organization in Living Systems"; see also Dupuy, *The Mechanization of the Mind*, ix, 120.

27. For von Foerster's own exposition of the tesseract, see "Cybernetics of Epistemology," 237–41. On the wider culture of ideas since the nineteenth century regarding the *spatial* construction of the fourth dimension, see Henderson, *The Fourth Dimension*. I am grateful to Professor Henderson for sharing with me her correspondence from von Foerster, a fan letter on the publication of her text cited above.

28. Von Foerster, "On Self-Organizing Systems," 13.

29. Ibid.

30. Ibid.

31. Von Foerster, "On Constructing a Reality," 216.

32. Luhmann, "The Control of Intransparency," 361.

33. On Luhmann's relation to von Foerster, see the apparatus of his work for pervasive references to numerous von Foerster articles. For instance, the very late, fairly short paper cited above, "The Control of Intransparency," mentions four von Foerster texts: "On Self-Organizing Systems" (in note 4); *Observing Systems* (in note 8); "Principles of Self-Organization" (in note 9); and the highly obscure 1948 text, von Foerster's calling card when first visiting the United States, *Das Gedächtnis* (in note 21). See also Baecker, "Knowledge and Ignorance."

34. "In the first quarter of [the twentieth] century physicists and cosmologists were forced to revise the basic notions that govern the natural sciences. . . . It was clear that the classical concept of an 'ultimate science,' that is an objective description of the world in which there are no subjects (a 'subjectless universe'), contains contradictions. To remove these one had to account for an 'observer' (that is at least for one subject): (i) Observations are not absolute but relative to an observer's point of

view (i.e. his coordinate system: Einstein); (ii) Observations affect the observed so as to obliterate the observer's hope for prediction (i.e. his uncertainty is absolute: Heisenberg)" (von Foerster, "Notes on an Epistemology for Living Things," 247).

35. Luhmann, "The Cognitive Program of Constructivism," 132.

36. Von Foerster, *Understanding Understanding*, 252. Original emphasis.

37. The problematics of communication under a second-order cybernetic constructivist regime, as those have been developed and variously resolved by von Foerster, Maturana, Varela, and Luhmann, are further addressed, in synthetic relation to C. S. Peirce, Jakob von Uexküll, and the conceptual program of biosemiotics, in Brier, "The Construction of Information and Communication."

38. Luhmann, *Social Systems*, 29.

39. Ibid.

40. Von Foerster, "On Constructing a Reality," 225.

41. Luhmann, "The Cognitive Program of Constructivism," 129.

42. Von Foerster, "On Constructing a Reality," 226.

43. Von Foerster, *Understanding Understanding*, 4.

44. Von Foerster, "On Constructing a Reality," 212.

45. Luhmann, *Social Systems*, 9.

46. Ibid.

47. Ibid. My emphasis.

48. On this point, see my *Posthuman Metamorphosis.*

49. Luhmann, *Social Systems*, 9.

50. Von Foerster, "On Constructing a Reality," 227.

The Early Days of Autopoiesis

FRANCISCO J. VARELA

In this memoir, the origin of the notion of autopoiesis is presented from the point of view of the basic intellectual background which gives its specificity and from the role of the major actors involved in its articulation and its setting in Chile in the late 1960s. Heinz von Foerster's role in this connection is exemplary, as he was an active, visionary, and supportive participant in this evolving conversation.

My homage to Heinz will be a story. He has been a mentor, friend, and inspiration for over thirty years, and there would be many stories to tell. But I think it is most appropriate to dwell on a particularly rich one: his role in the gestation and early days of the notion of autopoiesis. This is the first occasion in which I have publicly spoken about this period, and although I think it is important to go beyond the individual's roles, there is also the background of ideas and of social context that makes science alive.[1]

What does it mean when an idea like autopoiesis, in its strict sense, a theory of cellular organization, gains visibility and prominence beyond professional biology and becomes capable of affecting distant fields of knowledge? My answer is that ultimately we can only understand this phenomenon because the idea contains *a background of important historic sensibilities* with which it aligns and resonates. This background of tendencies does not appear strictly delineated but rather as a retrospective, because ideas, like history, are possibilities to be cultivated, not a result of some mechanical determinism. At this distance, autopoiesis holds a privileged place, in my opinion, for having clearly and explicitly announced a tendency which today is already a force in many areas of cultural inquiry.

The tendency to which I refer, stated briefly, is the disappearance of intellectual and social space which makes cognition a mentalist representation and the human being a rational agent. It is the disappearance of what Heidegger calls the period of the image of the world and what could also be referred to as Cartesianism. If autopoiesis has been influential it is because it was able to align itself with another project which focuses on the *interpretive capacity of the living being* and conceives of the human as an agent which doesn't discover the world but rather constitutes it. It is what we could call the ontological turn of modernity, which, toward the end of the twentieth century, is taking shape as a new space for social interaction and thought and which, undoubtedly, is progressively changing the face of science. In other words, autopoiesis is part of a picture much larger than biology, in which today it holds a privileged position. It is this syntony with a historical tendency, intuited more than known, which is the core of the early ideas on autopoiesis and whose development I hope to trace.[2]

The act of signing one's name to a text, more than claiming it as a personal possession, represents the placing of a milestone on a path. Ideas appear as movements of historical networks in which individuals are formed, rather than vice versa. Thus, Darwin already had Wallace waiting for him and Victorian England as the substratum; Einstein alone in his Swiss patent office had dialogues with Lorentz against the backdrop of the world of German physics at the end of the century; Crick was already familiar with the ideas of Rose and Pauling when he met Watson, and his attitude was that of Cambridge in the 1950s. *Mutatis mutandis*, the history of autopoiesis also emerges out of prior work and is nourished by a unique substratum. It was all of Chile that played a fundamental role in this story.

Writing this story is, I insist, making a fold in history where men and ideas live because we are points of accumulation among the social networks in which we live, rather than individual wills or characters. One cannot claim to draw together the density of actions and conversations that constitute us in a necessarily unidimensional personal account. I don't pretend that what I say here is an objective narrative. What I offer is, for the first time, my own tentative and open reading of how the notion of autopoiesis emerged and what has been its significance and development since. I have let everything I say mature over the years, and I believe it to be honest to the degree that I can take responsibility for being one of the direct participants in this creation, while maintaining an awareness that I cannot consider myself a holder of the truth.

To illuminate the background, I must begin with the roots of this story from my personal point of view. Paradoxically, only through recovering how the background appeared in the specificity of my perspective can I communicate to the reader the way in which this invention found its place on a broader horizon.

The Years of Incubation

I belong to a generation of Chilean scientists who had the privilege of being young during one of the most creative periods in the history of the Chilean scientific community, the decade of the 1960s. As a teenager, I had an early vocation for intellectual work, and the biological sciences seemed my undoubted destiny. Upon finishing secondary school in 1963, I opted for the Universidad Católica which announced an innovative undergraduate program in "Biological Sciences" following the third year of Medicine. As a medical student, I got to know the first researchers who fascinated me, people such as Luis Izquierdo, Juan Vial, Hector Croxato, and above all Joaquín Luco, who definitively infected me with a passion for neurobiology. Not far into my first year, I asked Vial if he would take me on as an apprentice in his cellular biology laboratory. He gave me the key to a little door to his laboratory, overlooking Calle Marcoleta, where I spent my free time staining myelin on nerve cross-sections.

Juan Vial also gave me good advice, including his recommendation, in 1965, that I move to the newly opened Department of Sciences at the University of Chile to continue my training. It was a crucial step, because I left the world of traditional careers in order to fully enter the universe of exclusive scientific training, until then unknown in Chile. In a few borrowed classrooms on the top floor of the School of Engineering, I found my place to grow: a small group of young people excited about research and pure science and researcher-professors who taught future scientists with passion.

Apprentice to a Neurobiologist

The last piece of advice Vial gave me was to find work with Humberto Maturana, who had just left the University of Chile's Medical School for the new Department of Sciences. On a beautiful day in April 1966, I went to see him in his laboratory in the basement of one of the sections of the new school on Calle Independencia. At that time Maturana was already an important researcher, known for his work on the physiology of vision in several classic papers he had written at Harvard and MIT before returning to Chile.[3] In Chile he continued to work on the physiology and anatomy of the retina in vertebrates.

To continue my apprenticeship in the trade, Humberto asked me to repeat experiments in electric recording on the optic tectum of the frog, which led me to investigate the problems of vision more deeply than I had ever done with a scientific problem. When I left the laboratory on Independencia to leave for the United States two years later, I had developed the ability to generate my first research ideas. One was a hypothesis about the role of time in the operation of the retina, which led to some experimental predictions which were the origin of my first scientific article.[4] Maturana's influence was one of the pillars he gave me during my years of apprenticeship in Chile, but it is important that I touch on at least two other influential currents which had and continue to have an enormous impact on my intellectual history. The first was philosophy and certain key readings I discovered during these years of training. The second was discovering the world of cybernetics and theoretical biology; in both areas Heinz's role was to become essential.

Philosophical Reflection

During my high school years my readings in philosophy were as passionate as they were random, mixing Aristotle, Ortega y Gasset, Sartre, and Papini. In search of a more systematic training, when I transferred to the Department of Sciences in 1966, I also enrolled in philosophy at the old Instituto Pedagógico and began to participate regularly in guided readings with Roberto Torreti in the Humanities Center at the School of Engineering. The Institute's grand ideological controversies didn't interest me as much as what I could discover thanks to the classes of Francisco Solar, which resonated with the German training of Torreti and which took form in the collections of the Center's library. There I discovered European phenomenology and began a reading, which continues to this day, of Husserl, Heidegger, and Merleau-Ponty. For the first time I seemed to find in these authors a preoccupation for the definition of the range of lived experience which I consider fundamental.

The second stunning discovery of these years was the social nature of science. I owe to Felix Schwartzman my early introduction to this world. In his course in the Department of Sciences, I came to know what until then was only known by a minority in Chile, the works of the French school in the history and philosophy of science: Alexandre Koyré (above all), Georges Canguilhem, and Gaston Bachelard. All of these authors express the counterintuitive conviction that scientific ideas are made and change abruptly, and not because of a lucky accumulation of "purely empirical evidence"; that they are sustained with images and ideas which are neither given nor immutable; and that each age is blind to the foundation of what it considers certain and evident. The general public became aware of all this through Thomas Kuhn's famous book,[5] which couldn't have existed without the groundwork of the French school, which Kuhn quotes with reverence. Barely nineteen, I was relieved forever of my position as naive apprentice. Schwartzman's guided readings on the mission of the scientist turned me into a critic of what I was receiving as my professional training.

Brain, Machines, and Mathematics

During that pioneering era, the Department of Sciences made few concessions when it came to training in mathematics. On my first day of class, without saying a word, the professor began to write: "Let E be a vector space. The axioms for E are. . . ." After the initial shock of getting up to speed, I discovered in mathematics a language and a way of thinking that fascinated me. It was at this ripe time that I first encountered Heinz as exponent of mixing mathematics with brain studies. Although I didn't meet him in person until 1968, he became quite immediately a figure of great importance for me. His papers circulated in the laboratory, with fascinating titles such as "Circuitry of Clues to Platonic Ideation."[6]

In these sources I first realized a long tradition which seeks to express the properties of biological phenomena beyond their material particularities. As we all know now, it is a way of thinking that had only appeared in the 1950s, more specifically with the publication of *Cybernetics*, by Norbert Wiener, and under the influence of another important person at MIT, Warren McCulloch,[7] whom Humberto had met in 1959–60 when he was working at MIT. Wiener, McCulloch, and von Foerster were *the* pioneers of the *conjunction* of epistemological reflection, experimental research, and mathematical modeling. Only many years later was I able to appreciate these early days of cybernetics and the major role Heinz played in them as editor to the Macy Conferences.[8]

Entry into Experimental Epistemology

Apprenticeship for the trade of neurobiologist wasn't the only thing going on in the laboratory. Humberto had entered a period of frank questioning of certain dominant ideas in neurobiology; discussion, reading, and debate were daily events, spurred on by the presence of Gabriela Uribe, a physician of clear epistemological leanings who was working with Maturana at that time. Those were times of search and discussion focusing on what seemed a dissatisfaction, an *anomaly*. A basic dissatisfaction was the notion of information as the key to understanding the brain and cognition; the idea didn't appear to play an explicit role in the biological process. Humberto's intuition was that living beings are, as he said in those days, "self-referred," and in some way the nervous system was capable of generating its own conditions of reference. It was a question of reformulating an orientation into an "experimental epistemology," a wonderful term introduced by McCulloch. Gabriela and Humberto had begun a study of certain chromatic effects similar to those described by E. Land in 1964, which were transformed into the topic around which a first attempt to reformulate visual perception as nonrepresentational was based.

The days of my training in Chile were coming to an end. The Biology Department offered to support me in obtaining a scholarship from Harvard University to do a doctorate. I began to wrap up my student life in Chile aware that I was leaving with a clear focus in experimental epistemology and with three living pillars in my imagination.

Harvard and the Crisis of 1968

I left for Harvard in a Braniff jet on January 2, 1968, reading a text by Koyré on Plato. I arrived in Cambridge in the midst of a snowstorm, with no place to live, far from speaking fluent English, and with the threatening knowledge that if I didn't get straight A's, my scholarship would be taken away. The first few months were hard, but once settled and getting to know my way around this new kingdom, I leaped head first into courses and seminars of all kinds: anthropology (studies on the natural ethology of primates were beginning); evolution (Stephen Jay Gould had just arrived at Harvard and was a sharp contrast to the classicist, Ernst Mayr); mathematics (the theory of dynamic nonlinear systems was discovered at this time); and philosophy and linguistics (Chomsky was the dominant figure along with Putnam and Quine). I found in Cambridge libraries until then only imagined, well stocked and open at all hours. I had the impression of having leapt into another galaxy, and I don't remember a single day in which I didn't feel like greedily absorbing everything at hand.

Long afterward I realized, with great surprise, that compared to most of my doctoral classmates, my interests and vision of science were frankly more heterodox and mature. Beyond that, I realized that talking with professors about epistemological problems, as I was accustomed to doing in Santiago, was not looked upon favorably. The reaction was the same when I attempted to find a way to cultivate my interests in theoretical biology. The MIT of 1968 had already disappeared, with McCulloch retired and no one to replace him. My only point of reference continued to be von Foerster, whom I visited several times at the Biological Computer Laboratory at the University of Illinois in Urbana, an active and productive center which he directed in those years. It was easy to see that my intellectual quest would have to be divided in two: the official and the private.

Officially, I was studying under Keith Porter, in whose laboratory I learned to work in cellular biology, and Torsten Wiesel, who not long thereafter received the Nobel Prize for his work on "information processing" in the visual cortex. I focused my interest on comparative aspects of vision and began work on the functional structure of the eyes in insects, which would become the subject of my dissertation. By early 1970 I had already published four articles on the topic, and my dissertation was accepted in April of 1970.

Unofficially, outside the laboratory, I found myself for the first time living in a world infinitely more vast than that of Santiago, with young people from other cultures, in a place where nationalities and races blended. As fate would have it, those were the years of the mythic events that marked my generation. What began in Paris on the night of May 10, 1968, corresponded to the movement in North America, centered around its opposition to the Vietnam War. The Kent State incident was followed by the first student strikes in which I took part. There were dramatic moments like the night the police forced us out of Harvard Yard. The Cambridge years were for me the discovery of my involvement as a member of society and the possibility of taking responsibility for changes in my social surroundings. It was a rediscovery of myself, far from my Latin American roots, which my friends from The Movement exalted in the form of the Cuban revolution. I was not only occupied with science, but also with the dream of a new Latin America belonging to our generation.

Having discovered myself to be a social and political animal accentuated the need to maintain a public silence regarding my true interests. Faithful to the idea of science as an activity that is made and created by jumps and bold innovations, like other members of my generation, I cultivated the intention to return to Chile to create a different science, in which the anomalies which had already appeared in Chile and were accentuated in the United States could be transformed into scientific practice. Creating my own original science seemed to me to define my obligation to my past and my roots.

I graduated as Doctor of Biology in June of 1970. Against the protests of my professors, I declined a post as a researcher at Harvard and another as Assistant Professor at another American university. I decided to accept a position offered to me by the Department of Sciences, justifiably interested in a return on the investment they had made in my training. I returned to Chile on September 2, 1970, and Allende's election two days later seemed to be my second and true graduation. At last the work could really begin, with key problems well defined, with the certainty of being as competent and well prepared as anyone on the world scientific scene, and within the context of working in an environment that had a future to build. Having provided the backdrop of the situation in

September 1970, I can return now to the specifics of the notion of autopoiesis and its gestation.

The Gestation of an Idea

Examining the Problem

The direct antecedent to the gestation of autopoiesis is the text that Maturana wrote in mid-1969, originally entitled "Neurophysiology of Cognition." Humberto had continued along his own line of questioning regarding the inadequacy of the ideas of information and representation to understand the biological system. He visited me on several occasions in Cambridge and, as in Santiago, we had long conversations. In the spring semester of 1969, Heinz invited him to come to the Biological Computer Laboratory for a few months, an opportunity which coincided with the international meeting of the Wenner-Gren Foundation on the subject of "Cognition: A Multiple View," a visionary title in light of the enormous development of what today are called the cognitive sciences but until then were not considered a scientific field.

Humberto prepared the text for this meeting, providing for the first time a clear and attractive expression of his matured ideas, in order to clarify what until then he alluded to as the self-referred nature of living beings and to definitively identify the notion of representation as the epistemological pivot which had to be changed. From his point of view, it was necessary to center attention on the internal linking of neuronal processes and to describe the nervous system as a "closed" system, as the text states. This article marks an important jump, and to this day I still believe that it was the indisputable beginning of a turn in a new direction. I remember having visited Humberto in Illinois and having discussed several difficult parts of the text while he was finishing it. The text appeared shortly thereafter, and the article opens with a paragraph thanking Heinz and me for the conversations we had on the topic.[9] Not long after that Humberto reworked the text into a more definitive version which came to be called "Biology of Cognition."

This text touches summarily on an idea that had been intriguing me for some time and that as an assistant in the cellular biology course taught by George Wald and James Watson at Harvard had appeared to me as a clear anomaly: one talked about the molecular constitution of the cell and used terms like self-maintenance, but no one, not even the two reunited Nobel Prize winners, knew what was meant by that. What was worse was that when I pushed the discussion in that direction during lunch, the habitual reaction was a typical, "Francisco, always getting into philosophy." My notes from that time include several attempts to examine the basic autonomy of the cellular process as the basis of the autonomy of life. Toward the end of 1969, Jean Piaget's *opus magnum*, entitled *Biologie et Connaissance*,[10] appeared in the window of Schoenhof's Foreign Books in Cambridge, in which he notes the clear need to reconsider biology on the basis of the autonomy of living systems, but Piaget's language and idiosyncrasies left me unsatisfied.

In his article, Humberto made the connection between the circular nature of neuronal processes and the fact that the organism is also a circular process of metabolic changes, as was illustrated with reference to a recent article by Commoner in *Science*, which discussed the new advances in the biochemistry of metabolism and its evolution. The question under examination then was: if we leave the organization of the nervous system to the side for the moment and focus on the autonomy of life in its cellular form, what can we say? This reflection on the circular nature of metabolism in living beings and its relation to cognitive operations, although barely filling a short page in the definitive version of "Biology of Cognition," would be a focal point from which the development of the idea of autopoiesis would be drawn.

Those were the final months of 1970, I was back in Chile, and the Biology Department approached me about taking on the introductory course in cellular biology for new students. Maturana and I were now colleagues in the Biology Department, with neighboring offices in the "transitional" (but still used) stalls in the new campus of the Department of Sciences on Calle Las Palmeras in Macul. Everything was in place to launch the exploration of the nature of the minimal organization of the living organism, and we didn't waste any time. In my notes the first mature outlines appear at the end of 1970, and toward the end of April 1971 appear more details along with a minimal model which would later be the subject of computer simulation. In May of 1971, the term autopoiesis appears in my notes as the result of the inspiration of our friend José M. Bulnes, who had just published a thesis on the *Quixote* in which he made use of the distinction between *praxis* and *poiesis*. A new word was appropriate because we wanted to designate something new. But the word only acquired power in association with the content our text assigned to it; its resonance reaches far beyond the mere charm of a neologism.

Those were months of almost constant work and discussion. Some of the ideas I tested with my students in the cellular biology course, others with colleagues in Chile. It was clear to us that we were embarking on a journey that was consciously revolutionary and anti-orthodox and that this valor had everything to do with the mood in Chile, where possibilities were unfolding into a collective creativity. The months that led to the development of autopoiesis are inseparable from Chile at that time.

During the winter of 1971, we knew that we were dealing with an important concept and we decided to put it in writing. A friend lent us his house on Cachagua beach, where we went twice between June and December. The days at the beach were divided between long walks and a monastic rhythm of writing, which Humberto usually began and which I took over later in the day. At the same time I began a first draft (which Humberto revised) of a shorter article which would set forth the principal ideas with the aid of a simulation of a minimal model (which we called the "Protobe"; more on this below). Around December 15 (again according to my notes of 1971), we had a complete version of a text in English called "Autopoiesis: The Organization of Living Systems." The typewritten version came to seventy-six pages, from which we made several dozen copies using the old blue ink mimeograph method. Although there were several later modifications, this text was to be published much later.

As has occurred often in the history of science, the creative dynamic between Maturana and myself resounded in an ascending spiral, to which a mature interlocutor contributed experience and previous consideration and a young scientist brought fresh perspectives and ideas. As is clear given the circumstances, the ideas did not emerge in one or two conversations, nor was it a simple question of making explicit what had already been said. What was in the background must be considered a *qualitative* leap. Such transitions are never simple, nor is it possible to retrace exactly how they came about, because there is always a blend of past and present, talents and weaknesses, imagination and inspiration. The mature concept of autopoiesis did have, as we have seen, clear roots, but between an idea and its roots exists a crucial jump. And just as Franklin's work is not the double helix of Watson/Crick, nor Einstein's that of Lorentz's special relativity, the key ingredients of autopoiesis cannot be reduced to the mature expression of the idea, as is easily seen comparing the published texts. This is a limpid example of what I had already learned from my French teachers, that science has discontinuities, that it doesn't function by progressive empirical accumulation, and that it is inseparable from its social and historical context.

One Idea and Two Texts

As is inevitable, understanding unravels over the course of time and in proportion to its effects. So it isn't surprising that the text we finished toward the end of 1971 wasn't accepted immediately. In fact it was sent to at least five publishers and journals, and without exception they considered it unpublishable. I remember in January of 1972, my ex-professor Porter invited me to visit the new Biology Department at the University of Colorado at Boulder, where I gave an enthusiastic talk entitled "Cells as Autopoietic Machines." The reception was cold and distant, as was that of my colleagues at Berkeley whom I visited around that same time.

The difficulties of finding a publisher, added to the political climate in Chile at the end of 1972, made me feel alienated from the international scientific world. At the same time, the enthusiastic reception of certain people whom I respected was of enormous value. The first to have a clear perception of the possibilities of the idea was naturally our friend Heinz in the United States, with whom we had been in constant communication and who came to Chile during those years. Another well-known cyberneticist and system theorist who reacted positively was Stafford Beer, who came to Chile on a regular basis. In fact, Fernando Flores had contracted with him on behalf of the government to implement a revolutionary system of communications and regulation of the Chilean economy inspired by the nervous system; the system came to be called Proyecto Cinco. Beer responded to what was set out in the text with such enthusiasm that we decided to ask him for a preface, which he agreed to write immediately. In January 1972, with a fresh copy of the manuscript, I was invited to Mexico by Ivan Illich, to his CIDOC center in Cuernavaca.[11] I gave him the manuscript the day I arrived, and I will never forget his reaction the following morning: "This is a classic text. You have managed to put autonomy at the center of science." Through Illich, the text made its way into the hands of the famous psychologist Erich Fromm, who invited me to his home retreat to discuss the new concept, which he immediately incorporated into the book he was writing at the time.[12] In Chile itself, Fernando Flores and other colleagues from Proyecto Cinco were also an attentive public to our way of thinking. With Flores we formed what would come to be a fruitful friendship, and many years later autopoiesis would figure among the important concepts he would use to develop his own ideas. It is hard to express what finding receptivity in people of this quality meant to me at the time.

Meanwhile the text continued to be rejected from a growing list of foreign publishers. So it was natural to address our own university press, and at the end of 1972 we signed a contract that included the translation of the text by Carmen Cienfuegos. *De máquinas y seres vivos: Una teoría de la organización biológica* was printed in April 1973. The original English text did not appear until 1980, when the idea had already acquired a certain popularity, in the prestigious Boston Studies on the Philosophy of Science series. This version contained an introduction signed by Maturana; the text, "Biology of Cognition"; Beer's preface; and the text in question, "Autopoiesis: The Organization of Living Systems."[13] According to what the editor tells me, this book has been the series' best-seller.

The brief article written in parallel to the longer text suffered a similar fate. As I mentioned above, in addition to a succinct presentation of the idea of autopoiesis, the intent of the article was to clarify the concept through a minimal case of autopoiesis. Toward the end of 1970 we had come to the conclusion that a simple case of autopoiesis would require two reactions: one of polymerization of membrane elements, the other, the "metabolic" generation of monomers. The latter had to be a reaction catalyzed by a third pre-existing element in the reaction. Once we had designed this reaction scheme, the next obvious step was to test a simulation of this minimal case (which soon came to be called the Protobe in our discussions) using cellular (or tessellated as they were called then) automata, introduced in the 1950s especially by John von Neumann. With the collaboration of Ricardo Uribe of the School of Engineering, the simulation rapidly provided the results our intuition had led us to expect: the spontaneous emergence in this artificial bi-dimensional world of units which self-distinguished by means of the formation of a "membrane" and which showed a capacity of self-repair. The paper was sent to several journals including *Science* and *Nature*, with results similar to those of the book: complete rejection. Heinz visited Chile in the winter of 1973 and helped us rewrite the text significantly. He took it back to the United States under his arm and sent it to the editor of the journal *BioSystems*, for which he was a member of the editorial board. The paper received some harsh commentary from reviewers but not long afterward was accepted and finally published in mid-1974.[14] It is important to mention this article here because it was the first publication on the idea of autopoiesis in English for an international public, which led the international community to take charge of the idea. In addition it anticipated what twenty years later would become the explosive field now called artificial life and cellular automata.

Heinz's visit in July of 1973 took place in the midst of the approaching storm which plunged us all into an atmosphere of permanent crisis, with desperate attempts to stabilize a country that was breaking in two. As a militant supporter of President Allende's government, after September 11, I found myself threatened. Military intelligence came to the department with lists of ex-party members, and on two occasions night patrols came looking for me at my house, where I no longer slept. I was dismissed from my post at the university on orders "from superiors." With my family I decided to sell everything and leave. The majority of my colleagues in the Department of Sciences also dispersed throughout the world. With the diaspora of the department's scientists ended a period of science in Chile, an important stage of my personal life, and with it the context which gave birth to the idea of autopoiesis. But naturally the idea would find new avatars, especially outside of Chile.

Coda

From my perspective in 1995, autopoiesis does not embody by itself a new vision of life and mind. Beside it appear other equally significant notions such as operational closure, enaction, natural drift, and phenomenological methodology.[15] The empirical references are consequently extended in new programs of detailed research, be they lymphocyte networks, the motion of insects, or cerebral imaging. It is a question of an edifice of new epistemological concepts and empirical results which have breadth and stand up to rigor. There have been twenty productive years during which the period of the formulation of autopoiesis marks, in retrospect, an important milestone, as should be evident to the reader who has been patient enough to follow me this far.

But if this slow, sustained construction, full of *corsi e ricorsi* as is all intellectual and scientific creation, today has scientific viability, it is because it forms part of a historic sensibility which autopoiesis intuited in 1970–71. As I said at the start, there are no personal creations without a context: that an idea has impact is a historical fact and not a personal adventure or a question of "being right." Autopoiesis continues to be a good example of alignment with something which only today appears more clearly configured in various fields of the human cultural endeavor and which I identify with the term ontological turn. That is, a progressive mutation of thought which ends a long dominance of the social space of Cartesianism and which opens up to the sharp consciousness that humankind and life are the conditions for the possibility of meaning and for the worlds in which we live. That knowing, doing, and living are not separate things and that reality and our transitory identity are partners in a constructive dance. This tendency I designate as an ontological turn is not a philosophical mode, but rather a reflection of the life of all things. We are entering a new period of fluidity and flexibility which drags with it the need to reflect on the way in which humans make the worlds they live in and do not find them already made as a permanent reference.

The occasion of writing this story twenty years later would be sadly wasted if I didn't manage to communicate the importance of expanding the horizon to consider the profoundly social and aesthetic nature from which this idea emerges, beyond science and biology and beyond the people named as authors. In this sense, the story of autopoiesis has not gone out of date and still can be read backwards in time with some profit. It is definitively a scientific invention, and all fields require actors who are sensitive to the anomalies which constantly surround us. These anomalies must be maintained in a state of suspension or cultivation while one can find an alternative expression which reformulates the anomaly as a central problem of life and knowledge. It is also an occasion for me to express my profound gratitude to Heinz, who was right there all along, a full participant in this moving conversation, and beyond that, a great teacher and dear friend.

Notes

From a festschrift for Heinz von Foerster guest-edited by Ranulph Glanville, this is a slightly emended and corrected version of Varela, "The Early Days of Autopoiesis: Heinz and Chile." Thanks to John Protevi for bringing it to our attention.—Eds.

1. The reader should be aware that this text has been adapted from a new preface for the twentieth-anniversary edition of the Spanish edition of Maturana and Varela, *De máquinas y seres vivos.* Thanks to Kirk Anderson for his help in the translation.
2. Syntony: A condition in which two oscillators have the same resonant frequency. —Eds.
3. See in particular the "classic": Maturana, Lettvin, McCulloch, and Pitts, "Anatomy and Physiology of Vision in the Frog."
4. Varela and Maturana, "Time Course of Excitation and Inhibition in the Vertebrate Retina."
5. Kuhn, *The Structure of Scientific Revolutions.*
6. For a selection of these and other articles, see von Foerster, *Observing Systems.* [See also von Foerster, "A Circuitry of Clues to Platonic Ideation." Varela wrote the introduction for *Observing Systems.* Currently, the most accessible selection of von Foerster's papers is *Understanding Understanding.*—Eds.]
7. A selection of his most important work is McCulloch, *Embodiments of the Mind.*
8. For an extraordinary account of the early days of cybernetics and the Macy Conferences, see Dupuy, *Aux sources des sciences cognitives.* [The revised translation is Dupuy, *The Mechanization of the Mind.*—Eds.]
9. Garvin, *Cognition.*
10. Piaget, *Biology and Knowledge.*

11. CIDOC: Centro Intercultural de Documentación (Intercultural Documentation Center).—Eds.

12. Fromm, *The Anatomy of Human Destructiveness.*

13. Maturana and Varela, *Autopoiesis and Cognition.*

14. Varela, Maturana, and Uribe, "Autopoiesis."

15. Varela, Thompson, and Rosch, *The Embodied Mind.*

Life and Mind

From Autopoiesis to Neurophenomenology

EVAN THOMPSON

Allow me to begin on a personal note. I first met Francisco Varela in the summer of 1977 at a conference called "Mind in Nature." The conference was organized by my father, William Irwin Thompson, and Gregory Bateson. It took place in Southampton, New York, at the Lindisfarne Association, an institute founded by my father, and was chaired by Bateson, who was then serving as Lindisfarne's scholar-in-residence.[1] I was not quite fifteen years old; Francisco was almost thirty-two. At that time Francisco was known within the circle of second-generation cybernetics and systems theory for his work with Maturana on autopoiesis and for his "calculus of self-reference."[2] But outside this circle he was known for an interview and a paper that had appeared about a year earlier in *CoEvolution Quarterly*, a widely read intellectual journal of the American counterculture in the 1970s.[3] The paper, called "Not One, Not Two," was a position paper on the mind–body relation, given at another conference, also involving Bateson and Heinz von Foerster, among others. I remember reading this paper, in which Francisco set forth some ideas about dualities and self-reference with application to the mind–body problem, and having the sense that it said something very important but without being able fully to understand it. I also remember listening to Francisco and the physicist David Finkelstein arguing about the relation between natural systems and logic and mathematics. Francisco was working on the algebraic foundations of self-reference and Finkelstein on quantum logic. Their debate was mesmerizing to me, even though I didn't have the knowledge or experience to follow it.

In preparing for this lecture today, I reread "Not One, Not Two," not having looked at it carefully for many years. What struck me this time are these words Francisco wrote at the end of the paper: "But what I see as an important ingredient of our discussion is the fact that a change in experience (being) is as necessary as change in understanding if any suturing the mind–body dualisms is to come about."[4] The dualism of concern to Francisco here was not the abstract, metaphysical dualism of mental and physical properties, but rather the dualism of mind as a scientific object versus mind as an experiencing subject. One of the most significant and exceptional aspects of Francisco's life and work, from this early paper to his last writings on his own illness and liver-transplant experience, is that he never lost sight of this point that the mind–body problem is not only a philosophical problem, or a scientific problem, but also a problem of direct experience.[5] The problem could be put this way. It's one thing to have a scientific representation of the mind as "enactive"—as embodied, emergent, dynamic, and relational; as not homuncular and skull-bound; and thus in a certain sense as insubstantial. But it's another thing to have a corresponding direct experience of this nature of the mind in one's own first-person case. In more phenomenological terms, it's one thing to have a scientific representation of the mind as participating in the "constitution" of its intentional objects; it's another thing to see such constitution at work in one's own lived experience. Francisco believed, like phenomenologists and also Buddhists, that this kind of direct experience is possible. He also thought that unless science and philosophy make room for this kind of experience, we will never be able to deal effectively with the mind–body problem but will instead fall prey to one or another extreme view—either denying experience in favor of theoretical constructions or denying scientific insight in favor of naive and uncritical experience.

Ten years later Francisco and I worked hard on developing these ideas when we began writing our book, *The Embodied Mind*, in 1986. If I may be bold, I think that although the ideas about embodied cognition in this book have been widely acknowledged and assimilated by the field, the book's central theme has yet to be fully absorbed. That theme is the need for back-and-forth circulation between scientific research on the mind and disciplined phenomenologies of lived experience. Without such circulation, the danger for the scientist and philosopher is nihilism, by which I mean the inability to stop experiencing things and believing in them in a way one's theory says is an illusion. Theoretical ideas like "being no one" (that there are no such things as selves but only neural self-models) or that consciousness is the brain's "user illusion" bear witness to this predicament.[6] An appreciation of what Francisco and I called the "fundamental circularity" of science and experience reminds us that such models of consciousness are objectifications that presuppose, on an empirical level, the particular subjectivities of the scientists who author them but also, on a transcendental level, the intentionality of consciousness as an *a priori* openness to reality, by virtue of which we are able to have any comprehension of anything at all. Experience is thus, in a certain sense, irreducible.

Let me jump ahead another ten years to Francisco's 1996 paper on neurophenomenology. Here the idea that the mind–body problem is also a problem of experience is articulated pragmatically in relation to neuroscience and the so-called "hard problem" of consciousness. The hard problem of consciousness is the problem of how and why physiological processes give rise to experience. It's one thing to be able to establish correlations between consciousness and brain activity; it's another thing to have an account that explains how and why certain physiological processes suffice for consciousness. At present, we not only lack such an account, but we are also unsure about the form it would need to have in order to overcome the conceptual gap between subjective experience and the brain. In proposing neurophenomenology as a "methodological remedy" for the hard problem, Francisco's insight was that no purely third-person, theoretical proposal or model would suffice to overcome this gap. "In all functionalistic accounts," he wrote, "what is missing is not the coherent nature of the explanation but its alienation from human life. Only putting human life back in will erase that absence; not some 'extra ingredient' or profound 'theoretical fix.'"[7] "Putting human life back in" means, among other things, expanding neuroscience to include original phenomenological investigations of experience. In this way, "the experiential pole enters directly into the formulation of the complete account," rather than being merely the referent of yet another abstract functionalist model.[8] But if experience is to play a central role in this way, then it has to be mobilized according to a rigorous phenomenology. Pragmatically, this means that the neuroscience of consciousness needs to incorporate disciplined, first-person investigations of experience, as illustrated in a preliminary way by one of Francisco's last experimental studies, which used original first-person data to guide the study of brain dynamics.[9] On such an approach, phenomenology is "not a convenient stop on our way to a real explanation, but an active participant in its own right."[10] In Francisco's words, "Disciplined first-person accounts should be an integral element of the validation of a neurobiological proposal, and not merely coincidental or heuristic information."[11]

In addition to this new methodological approach, neurophenomenology is also informed by an autopoietic conception of life; an enactive conception of mind; and a phenomenological conception of intentionality, subjectivity, and the lived body. These link neurophenomenology to what Francisco called "renewed ontologies" of mind and life.[12] This idea of renewed ontologies is what I want to talk about today.

Life Beyond the Hard Problem

My first step is to recast the terms in which the hard problem of consciousness is usually stated. Consider Thomas Nagel's classic formulation of the hard problem: "If mental processes are physical processes, then there is something it is like, intrinsically, to undergo certain physical processes. What it is for such a thing to be the case remains a mystery."[13]

Nagel's point is the now familiar one that we don't understand how an objective physical process could be sufficient for or constitutive of the subjective character of a conscious mental process. But stating the problem this way embeds it within the Cartesian framework of the "mental" versus the "physical," and this framework actually promotes the explanatory gap and so is incapable of resolving it. What we need instead is a framework that doesn't set "mental" and "physical" in opposition to each other or reduce one to the other ("not one, not two"). We need to focus on a kind of phenomenon that is already beyond this gap. *Life* or *living being* is precisely this kind of phenomenon. For biology, living being is *living organisms*; for phenomenology, it is *living subjectivity*. Where these two meet is in what phenomenologists call the *lived body*. What we need, and what neurophenomenology aims for, is an account of the lived body that integrates biology and phenomenology and so goes "beyond the gap."[14] What happens if we substitute "body" for "physical" in Nagel's statement?

> If mental processes are bodily processes, then there is something it is like, intrinsically, to undergo certain bodily processes.

Does this substitution make any difference? If there belongs to certain bodily processes something it is like to undergo them, then those bodily processes are experiences. They have a subjective or first-person character, which they could not lack without ceasing to be experiences. They are *feelings*, in the broad sense William James had in mind when he used the word "feeling" "to designate all states of consciousness merely as such" and that Damasio has revived by describing feelings as "bearing witness to life within our minds."[15] The problem of what it is for mental processes to be also bodily processes is thus in large part the problem of *what it is for subjectivity and feeling to be a bodily phenomenon.* In phenomenological language: What is it for a physical living body (*Körper/ leiblicher Körper*) to be also a lived body (*Leib/körperlicher Leib*)?

It's tempting to call this problem the *body–body problem*. I offer it as a "radical embodiment" reformulation of the hard problem.[16] In putting the problem this way, I am relying on the phenomenological distinction between the body as a material thing (*Körper*) and the body as a living and feeling being (*Leib*). This distinction is between two modes of appearance of one and the same body, not between two bodies or two properties (in the property-dualist sense). Hence the explanatory gap is now between two types within one typology of embodiment or living being, not between two opposed and reified ontologies ("mental" and "physical"). Furthermore, this gap is no longer absolute because in order to state it we need to make common reference on both sides to *life* or *living being*.

These two points are philosophically nontrivial. In the hard problem as classically conceived, the gap is absolute because there is and can be no conceptual unity to the mental and the physical, consciousness and the brain. Consciousness is equated with qualia, which are supposed to be phenomenal properties that resist functional analysis, while the body is equated with structure and function, with mechanism.[17] Given these equivalences, one must either mechanize consciousness in order to reduce it to a brain state or be a property dualist. This way of dividing up the universe is thoroughly Cartesian. Although physicalist philosophy of mind today rejects Descartes's substance dualism, it maintains both the underlying conceptual separation of mind and life and the equation of life with mere mechanism.[18]

For neurophenomenology, by contrast, the guiding issue isn't the contrived problem of how to derive a subjectivist concept of consciousness from an objectivist concept of the body. Instead, it's to understand the *emergence of living subjectivity from living being, including the reciprocal shaping of living being by living subjectivity*. It's this issue of *emergence* that neurophenomenology addresses, not the Cartesian version of the hard problem.

The Strong Continuity of Life and Mind

Implicit in this step of recasting the terms of the hard problem is the idea of a *strong continuity* of life and mind. One way to put this idea is that life and mind share a common pattern or organization, and the organizational properties characteristic of mind are an enriched version of those fundamental to life.[19] Mind is life-like, and life is mind-like. But a simpler and more provocative formulation is this one: *Living is cognition*.

This proposition comes from Maturana and Varela's theory of autopoiesis.[20] Some have taken the "is" in this proposition as the "is" of identity (living = cognition),[21] others as the "is" of predication or class inclusion (all life is cognitive).[22] The origins of the proposition go back to Maturana's 1970 paper, "Biology of Cognition." There he used the concept of cognition widely to mean the operation of any living system in the domain of interactions specified by its circular and self-referential organization. Cognition is effective conduct in this domain of interactions, not the representation of an independent environment. In Maturana's words, "*Living systems are cognitive systems, and living as a process is a process of cognition.* This statement is valid for all organisms, with and without a nervous system."[23]

Francisco later came to prefer a different way of explicating the "living is cognition" proposition: *Living is sense-making.* Consider motile bacteria swimming uphill in a food gradient of sugar. The cells tumble about until they hit on an orientation that increases their exposure to sugar, at which point they swim forward, up-gradient, toward the zone of greatest sugar concentration. This behavior happens because the bacteria are able to sense chemically the concentration of sugar in their local environment through molecular receptors in their membranes, and they are able to move forward by rotating their flagella in coordination like a propeller. These bacteria are of course autopoietic. They also embody a dynamic sensorimotor loop: the way they move (tumbling or swimming forward) depends on what they sense, and what they sense depends on how they move. Moreover, the sensorimotor loop both expresses and is subordinated to the system's autonomy, to the maintenance of its autopoiesis. As a result, every sensorimotor interaction and every discriminable feature of the environment embodies or reflects the bacterial perspective. For instance, although sucrose is a real and present condition of the physicochemical environment, its status as food is not. That sucrose is a nutrient isn't intrinsic to the structure of the sucrose molecule; it's a relational feature, linked to the bacterium's metabolism. Sucrose has significance or value as food, but only in the milieu that the organism itself brings into existence. Francisco summarized this idea by saying that thanks to the organism's autonomy, its world or niche has a "surplus of significance" compared with the physicochemical environment.[24] Living isn't simply a cognitive process; it's also an *emotive* process of sense-making, of bringing signification and value into existence. In this way the world becomes a place of *valence*, of attraction and repulsion, approach or escape. This idea can be depicted in the diagram below.[25]

Using this representation, I would like to expand the proposition "living is sense-making" in the following way:

1. *Life = autopoiesis.* By this I mean the thesis that the three criteria of autopoiesis—(i) a boundary, containing (ii) a molecular reaction network,

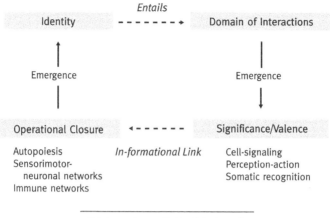

Entails

| Identity | - - - - - - → | Domain of Interactions |

Emergence ↑   Emergence ↓

| Operational Closure | ◄ - - - - - - | Significance/Valence |

Autopoiesis   *In-formational Link*   Cell-signaling
Sensorimotor-   Perception-action
  neuronal networks   Somatic recognition
Immune networks

"Living = Sense-Making" (redrawn).

that (iii) produces and regenerates itself and the boundary—are necessary and sufficient for the organization of minimal life.

2. *Autopoiesis entails emergence of a self.* A physical autopoietic system, by virtue of its operational closure, gives rise to an individual or self in the form of a living body, an organism.

3. *Emergence of a self entails emergence of a world.* The emergence of a self is also by necessity the emergence of a correlative domain of interactions proper to that self, an *Umwelt.*

4. *Emergence of self and world = sense-making.* The organism's world is the sense it makes of the environment. This world is a place of significance and valence, as a result of the global action of the organism.

5. *Sense-making = cognition (perception/action).* Sense-making is tantamount to cognition, in the minimal sense of viable sensorimotor conduct. Such conduct is oriented toward and subject to signification and valence. Signification and valence do not preexist "out there" but are enacted or constituted by the living being. Living entails sense-making, which equals cognition.

At this point you may want to object that the proposition "life is cognition" conflates *cognition* with *adaptation.* Margaret Boden makes this charge.[26] She thinks it would be better to use the term "cognition" more strictly to avoid the implication that autopoiesis necessarily involves cognition.

I disagree. We need to ask what exactly is meant by "adaptation." For Neo-Darwinians evolution involves the *optimization of adaptation* through natural selection. But from the autopoietic perspective, evolution involves simply the *conservation of adaptation*: as long as a living being doesn't disintegrate but maintains its autopoietic integrity, it is *ipso facto* adapted because its mode of sense-making continues to be viable. From this point of view, adaptation is an invariant background condition of all life. "Cognition," on the other hand, in the present context means the sense-making activity of living, which underlies the conservation of adaptation—no sense-making, no living, no conservation of adaptation. Notice that this way of thinking about cognition rests on an explicit hypothesis about the natural roots of intentionality: *intentionality arises from the operational closure of an autonomous system, whose paradigm and minimal case is an autopoietic system.* This hypothesis also amounts to a proposal about how to connect the phenomenological conception of intentionality to biology and complex systems theory.

At the other end of the spectrum from Boden, the biologist Lynn Margulis speaks of "microbial consciousness" and suggests that the "conscious cell" is the evolutionary antecedent of animal consciousness and the nervous system.[27] And the phenomenologist Maxine Sheets-Johnstone, in *The Primacy of Movement*, argues that bacteria aren't simply cognitive but embody a rudimentary kind of corporeal consciousness.[28]

You might be tempted to dismiss this idea of cellular consciousness out of hand. But let's consider the idea for a moment. "Consciousness" can have many meanings, but the one most relevant here is *sentience*, the feeling of being alive and exercising effort in movement. Maine de Biran wrote of *le sentiment de l'existence*. Both Damasio and Panksepp talk about a primitive *feeling of self*.[29] Phenomenologists, from Patocka to Sheets-Johnstone to Barbaras, call attention to the importance of *movement* for understanding the intentionality of consciousness.[30] Margulis, like Rodolfo Llinás, describes conscious thought as mental movement.[31] She believes that as brain activity it derives from ancient motile bacteria, which have left their evolutionary stamp on the cellular architecture and communication of neurons. One might summarize these threads by saying that consciousness as sentience is a kind of *primitively self-aware liveliness or animation of the body*. Does this emerge with life itself, with the very first living bodies—namely, bacterial cells? Hans Jonas poses the problem clearly:

> At which point . . . in the enormous spectrum of life are we justified in drawing a line, attributing a "zero" of inwardness to the far side and an initial "one" to the side nearer to us? Where else but at the very beginning of life can the beginning of inwardness be located?[32]

Whether we give this inwardness the name of feeling, receptiveness or response to stimuli, volition, or something else—it harbors, in some degree of "awareness," the absolute interest of the organism in its own being and continuation.[33]

This "absolute interest of the organism in its own being and continuation" is what Spinoza called *conatus*, the "concern" to exist, to carry on being, that belongs to life. Jonas observes that Spinoza, with the knowledge of his time, didn't realize that this concern "can only operate as a movement that goes constantly *beyond* the given state of things" and so is never a matter of mere preservation.[34] We, however, can observe that Jonas, with the knowledge of his time, didn't realize that this self-transcending movement is a natural consequence of a certain kind of self-organizing, morphodynamic system—namely, an autopoietic one. The self-transcending movement of life is none other than metabolism, and metabolism is none other than the biochemical instantiation of the autopoietic organization. That organization must remain invariant—otherwise the organism dies—but the only way autopoiesis can stay in place is through the incessant material flux of metabolism. In other words, the operational *closure* of autopoiesis demands that the organism be an *open system*. Jonas called this condition the "needful freedom" of the organism. The organism is never bound to its material composition at any given instant, but by the same token it has to change because stasis means death.

Coming back to the question about consciousness, I think that life's sense-making is a manifestation of the organism's autonomy and coupling, but not necessarily of consciousness. In support of my preference for this view I would appeal to the following considerations. First, being "phenomenally conscious" of something would seem to entail being able to form intentions to act in relation to it.[35] It's hard to make sense of the idea of being conscious of something, in the sense of subjectively experiencing it, while having no intentional access to it whatsoever. But there seems no reason to think that autopoietic selfhood of the minimal cellular sort involves any kind of intentional access on the part of the organism to its sense-making. Second, it seems unlikely that minimal autopoietic selfhood involves phenomenal selfhood or subjectivity, in the phenomenological sense of a pre-reflective self-awareness constitutive of a phenomenal first-person perspective.[36] Rather, this would seem to require the reflexive elaboration and interpretation of life processes provided by the nervous system. Finally, it's important to situate consciousness in relation to dynamic, unconscious processes of life regulation, and this becomes difficult if one projects consciousness down to the cellular level.

Teleology and "Autopoietic Machines"

A number of things I've said so far suggest that living beings are in some sense teleological: organisms have an interest in their own being and continuation; they realize a dynamic impulse to carry on being; they are always impelled beyond their present condition—these are teleological modes of description. "Living is sense-making" also sounds like a teleological description because it characterizes the organism as oriented toward the sense it makes of its environment. "Sense-making" is reminiscent of the phenomenological notion of intentionality, which signifies not a static representational "aboutness" but rather an act of intending, a purposive striving focused on finding satisfaction in further cognitive acquisitions and experience.[37] Behind this concept of intentionality we can see the metaphor or kinaesthetic image schema (in Lakoff and Johnson's sense) of self-generated and goal-directed movement, the motility of life.

Yet how are we to understand this suggestion of teleology in relation to the theory of autopoiesis, which in its original formulation was mechanistic and anti-teleological? Maturana and Varela explicitly identified living systems with machines and denied that living systems are teleological: "Living systems, as physical autopoietic machines, are purposeless systems."[38] By "machine" they clearly did not mean an artifact. They meant any system whose operation is determined by its relational organization and the way that organization is structurally realized.[39] Autopoietic systems maintain their own organization constant through material change and thus are homeostatic (or homeodynamic) systems of a special sort.[40]

At this point we need to ask whether having a relational organization is sufficient for being a machine. We can also wonder about the notion of emergence in this context. The work Francisco and I did on emergence and whole-system causation would seem to conflict with his view that autopoiesis can be realized in a cellular automaton.[41] In a cellular automaton, there is arguably no genuine emergence and system causation because every unit is local and the global pattern is in the eye of the observer. We argued, however, that in real living systems, such as a cell or large-scale neural assembly, there is emergence and circular causality, such that the system moves as a whole and constrains the states of its components.

At this juncture, I think it may be useful to draw from another line of work in theoretical biology, the work of Robert Rosen. Rosen and Francisco share many ideas, although oddly they never mention each other in their writings.[42] Rosen's dictum is that organisms are different from machines because they are "closed to efficient causation."[43] In an organism, but not in a machine, every efficient cause is produced inside the organism. More abstractly stated, Rosen argues that in a relational model of an organism every function (in the mathematical sense of a mapping) is entailed by another function within the model, whereas in a relational model of a machine this closure doesn't obtain, and one has to go outside the system and appeal to the environment. As Rosen puts it, there's an "impoverishment of entailment" in a machine compared with an organism.[44] In Francisco's language, this difference corresponds to the difference between an autonomous system with operational closure and a heteronomous system defined by outside control.[45] But Rosen also argues that closure and maximal entailment in an organism can't be simulated by a Turing machine.[46] More precisely, he shows that a certain class of relational models called *Metabolism-Repair* systems or (M,R) systems, in which every function is entailed by another function inside the system, aren't Turing-computable. On this basis, he argues that any material realization of an (M,R) system, such as a cell, can't be a mechanism or machine. This raises the question of what the relation is between Rosen's (M,R) systems and autopoietic systems. In a recent article, Letelier, Marín, and Mpodozis argue that *autopoietic systems are a subset of Rosen's* (M,R) *systems*: every autopoietic system is operationally equivalent to an (M,R) system (but not conversely, because a generic [M,R] system lacks the autopoietic property of generating its own boundary and internal topology).[47] It would seem to follow that *autopoietic systems are not Turing-computable* and that a physical autopoietic system—an organism or living being—is not a machine (at least according to one abstract and powerful concept of mechanism).

If Rosen is right about life being noncomputable, then this result is an important challenge to the original placement of autopoiesis in the category of cybernetic mechanism. It also challenges the hypothesis that autopoiesis can be captured by cellular automata models and allows for a stronger notion of emergence than the emergence we see in cellular automata. Emergence is present when there is no way to analyze a system into preexisting parts and resultant whole. Maturana and Varela, and Rosen, in different ways argued that this sort of analysis or "fractionation" fails in the face of the organism's self-referential organization. Here, part and whole are completely interdependent: an emergent whole is produced by a continuous interaction of its parts, but these parts cannot be characterized independently from the whole.

We can now return to the issue of teleology. Francisco, in his articles up to the early 1990s, continued to resist the idea that autopoiesis involved anything teleological.[48] But in one of his last essays he changed his mind. This essay, written with Andreas Weber, concerns autopoiesis and the problem of teleology and the organism from Kant's *Critique of Judgement*.[49] There Francisco argues that *teleology arises from autopoiesis* and is none other than the organism's sense-making. Yet strangely he doesn't even mention, let alone discuss, the change from his earlier to later view, nor the reasons for the change. Nor does he comment on his earlier acceptance of the life-as-machine notion—another striking omission considering that one of Kant's main points was that organisms are "natural purposes" by virtue of being self-organizing and must be judged to be fundamentally different from machines.

Francisco's change of view reflects his immersion in phenomenology at the end of his life. Even his later articles disavowing teleology were written prior to this neurophenomenological phase.[50] The change of view also reflects his deeper study of traditions of biological thought influenced by Kant. The type of teleology Maturana and Varela criticized in *Autopoiesis and Cognition* was *teleonomy* or Neo-Darwinian functionalistic explanation. But the type of teleology Francisco later discussed is Kant's idea that the organism is a "natural purpose" because it is a self-organizing being.[51] Francisco came to think that autopoiesis provided a naturalistic way of grounding, reformulating, and advancing the Kantian view of living beings as teleological, in a way Kant thought was impossible.[52]

Francisco never tried to reconcile his earlier and later statements about teleology, but I think it might be possible to do so in the following way. The main point he and Maturana insisted upon in *Autopoiesis and Cognition* is that teleology does not belong to the *autopoietic organization*. This point remains valid: in setting out the conditions for a self-producing organization in the molecular domain, no reference is made to notions such as "end," "purpose," "goal," or "function." On the other hand, the main point of the later revision is that teleology is none other than sense-making. Sense-making is not a feature of the autopoietic *organization* but rather of the *coupling* of a concrete autopoietic system and its environment. In other words, teleology is not an intrinsic organizational property but an emergent relational one that belongs to a concrete autopoietic system interacting with its environment. Let me try to indicate where these reflections seem to lead. If living beings are not reducible to algorithmic mechanism and if teleology is an emergent relational property, not an intrinsic organizational one, then we are faced with the prospect of a new kind of biological naturalism beyond the classical opposition of mechanism and teleology. Francisco's intuition was that such a naturalism would be able to offer strong bridges to phenomenology, but also that phenomenology could contribute to its formulation. Thus naturalizing phenomenology, for Francisco, always implied a corresponding phenomenological reconceptualization of nature.

In preparing this talk, I was struck by the thought that maybe one reason Francisco revised his view about teleology, though perhaps not a consciously articulated one, was his immersion in the life process of his own chronic and terminal illness. Francisco experienced firsthand, in an intense and singular way, life's sense-making. He realized it through his experience of his own living being, as it suffered the changing anti-viral treatments, the liver transplant and its "offering" of life,[53] the chemotherapy, the fatigue of sickness, and his scientific and phenomenological curiosity about living and dying. Using a Freudian idiom, Francisco called this curiosity his "epistemo-philia." I would add that his epistemo-philia was unique, in its embodiment of Buddhist mental presence, mathematical insight, phenomenological intuition, and an exceptional biologist's "feeling for the organism."[54] Francisco's revisiting the problem of teleology reflects his deep insight that *the mind–body problem is first and foremost a problem of lived experience.* As he and Weber wrote in this last article on life and teleology, commenting on Kant and Jonas,

> It is actually by experience of our teleology—our wish to exist further on as a subject, not our imputation of purposes on objects—that teleology becomes a real rather than an intellectual principle. . . . Before being scientists we are first living beings, and as such we have the evidence of intrinsic teleology in us. And, in observing other creatures struggling to continue their existence—starting from simple bacteria that actively swim away from a chemical repellent—we can, by our own evidence, understand teleology as the governing force of the realm of the living. Theories about the living can only be conceived from the fragile and concerned perspective of the living itself: [and then quoting Jonas] ". . . life can only be known by life."[55]

Life Can Only Be Known by Life

To close this talk, I would like to comment on this proposition that life can only be known by life. The claim is a transcendental one in a Kantian and Husserlian sense: it's about the conditions for the possibility of knowing life, given that we do actually have biological knowledge. Consider the question, how is it that we are able to recognize or comprehend the form or dynamic pattern of autopoiesis in the first place? Would this pattern be recognizable at all from some ideal objective standpoint? Or is it rather that we're able to recognize this pattern only because it resembles the form of our own bodily selfhood, which we know firsthand? Here, in brief, is the phenomenologist's answer: (1) An adequate account of certain observable phenomena requires the concepts *organism* (in the original Kantian sense of a self-organizing whole) and *autopoiesis*. (2) The source of the meaning of these concepts is the *lived body*—our first-person, lived experience of our own animate, bodily existence. (3) These concepts and the biological accounts in which they figure aren't derivable, even in principle, from some observer-independent, nonindexical, objective, physico-functional description (according to the physicalist myth of science). As Jonas puts it, no disembodied and purely intellectual mind, like Laplace's divine mathematician, would be able to comprehend the *form* of the organism simply from a complete knowledge of the microphysical state of things. To make the link from matter to life and mind, from physics to biology, one needs concepts like *organism* and *autopoiesis*, but such concepts are available only to an embodied mind with firsthand experience of its own living body. In Merleau-Ponty's words, "*Je ne puis comprendre la fonction du corps vivant qu'en l'accomplissant moi-même et dans la mesure où je suis un corps qui se lève vers le monde*" (I cannot understand the function of the living body except by enacting it myself, and except in so far as I am a body which rises toward the world).[56]

Francisco said that the "basic ground" of neurophenomenology is the "irreducible nature of conscious experience." "Lived experience," he wrote, is "where we start from and where we all must link back to, like a guiding thread."[57] Let us be clear about what this means. Experience is irreducible not because it possesses metaphysically peculiar "properties" that can't be squeezed into some reified, physicalist model of the universe, after the fashion of contemporary property dualism. It's irreducible because of its ineliminable transcendental character: lived experience is always already presupposed by any statement, model, or theory, and the lived body is an *a priori* invariant of lived experience. Experience is *die Unhintergehbarkeit*—the "ungobehindable." There is no dualism or idealism here: the transcendental lived body is no other than the empirical living body; it's simply that body *re-membered* in a certain way—namely, as where we start from and where we must all link back to, like a guiding thread.

I began this talk on a personal note, and I would like to end it that way too. The first conversation I ever had with Francisco was while we drove together with my father from New York City to Southampton to the "Mind in Nature" conference in 1977. Not long before I had discovered the writings of Borges, which I proceeded to devour in the way only an adolescent mind can. Somehow Francisco and I fell into a conversation about literature, and I declared my enthusiasm for Borges. Francisco preferred Neruda. About a year later Francisco gave me a copy of the English translation of Neruda's *Memoirs*, inscribed, "To Evan Thompson, with love and friendship, Francisco, September 1978," which

I still have to this day. On the first page of this book, Neruda writes, "Perhaps I didn't live just in my self, perhaps I lived the lives of others. . . . My life is a life put together from all those lives: the lives of the poet." These words express a sentiment that Francisco's thought and life echo in so many ways. Speaking for myself, my talk today is an expression of my deep gratitude for the participation of Francisco's life in mine.

Notes

This text was presented to the conference "De l'autopoièse à la neurophénoménologie: Un hommage à Francisco Varela" (From autopoiesis to neurophenomenology: A tribute to Francisco Varela), June 18–20, 2004, at the Sorbonne in Paris. The text was written to be read aloud, and I have resisted the urge to alter it so that it can remain true to its inspiration and purpose.

1. The conference ran August 24–31, 1977. The participants were Lewis Balamuth, Gregory Bateson, Mary Catherine Bateson, David Finkelstein, David Fox, William Irwin Thompson, Francisco Varela, and Arthur Young. It took place while Bateson was working on the manuscript of his last book, *Mind and Nature: A Necessary Unity.*

2. Varela, Maturana, and Uribe, "Autopoiesis"; Maturana and Varela, *Autopoietic Systems*; Varela, "A Calculus for Self-Reference."

3. Varela, "On Observing Natural Systems" and "Not One, Not Two."

4. Varela, "Not One, Not Two," 67.

5. Varela, "Intimate Distances."

6. Metzinger, *Being No One*; Dennett, *Consciousness Explained.*

7. Varela, "Neurophenomenology," 345.

8. Ibid.

9. Lutz, Lachaux, Martinerie, and Varela, "Guiding the Study of Brain Dynamics by Using First-Person Data."

10. Varela, "Neurophenomenology," 344.

11. Ibid.

12. Varela, "The Naturalization of Phenomenology as the Transcendence of Nature."

13. Nagel, "What Is It Like to Be a Bat?" 175.

14. Roy, Petitot, Pachoud, and Varela, "Beyond the Gap."

15. James, *The Principles of Psychology*, 185; Damasio, *Looking for Spinoza*, 140.

16. Thompson and Varela, "Radical Embodiment"; Thompson, *Mind in Life.*

17. See Chalmers, *The Conscious Mind* and "Moving Forward on the Problem of Consciousness," and Kim, *Mind in a Physical World.*

18. This is clearly evident, for instance, in the widespread view that there is no hard problem of life because life is nothing but structure and function, whereas there is for consciousness because physical structure and function logically underdetermine phenomenal consciousness. See Chalmers, *The Conscious Mind*, 106–7, 169, and "Moving Forward on the Problem of Consciousness," 5–6.

19. Godfrey-Smith, "Spencer and Dewey on Life and Mind," 230; see also Wheeler, "Cognition's Coming Home."

20. Maturana and Varela, *Autopoiesis and Cognition.*

21. Stewart, "Life = Cognition" and "Cognition = Life."

22. Bourgine and Stewart, "Autopoiesis and Cognition;" Bitbol and Luisi, "Autopoiesis with or without Cognition."

23. Maturana, "Biology of Cognition," 13. Original emphasis.

24. Varela, "Organism" and "Patterns of Life." The second article was written for a conference in 1992.

25. The source of the diagram is Varela, "Patterns of Life."

26. Boden, "Autopoiesis and Life," 40.

27. Margulis, "The Conscious Cell."

28. Sheets-Johnstone, *The Primacy of Movement,* 52, 73.

29. Damasio, *The Feeling of What Happens*; Panksepp, "The Periconscious Substrates of Consciousness."

30. Patočka, *Body, Community, Language, World*; Sheets-Johnstone, *The Primacy of Movement*; Barbaras, "The Movement of the Living."

31. Margulis, "The Conscious Cell"; Llinás, *I of the Vortex.*

32. Jonas, *Mortality and Morality,* 63; see also Jonas, *The Phenomenon of Life,* 57, 58.

33. Jonas, *Mortality and Morality,* 69; see also Jonas, *The Phenomenon of Life,* 84.

34. Jonas, "Biological Foundations of Individuality," 243.

35. Hurley, *Consciousness in Action,* 149–50.

36. Zahavi, *Self-Awareness and Alterity.*

37. See Held, "Husserl's Phenomenological Method," 14.

38. Maturana and Varela, *Autopoiesis and Cognition,* 86.

39. Ibid., 75, 77. Varela says in a footnote: "In this book 'machines' and 'systems' are used interchangeably. They obviously carry different connotations, but the differences are inessential, for my purpose, except in seeing the relation between the history of biological mechanism and the modern tendency for systemic analysis. Machines *and* systems point to the characterization of a class of unities in terms of their organization" (*Principles of Biological Autonomy,* 7).

40. Maturana and Varela, *Autopoiesis and Cognition,* 78, 79.

41. Thompson and Varela, "Radical Embodiment."

42. Like Francisco, Rosen died not long ago (in 1998) and too young (sixty-four) from complications of an illness. He lived in my home country, Canada, and I regret that I didn't study his work in time to talk to him and Francisco about their ideas. A worthwhile scientific and epistemological project would be to assess their theories in relation to one another, and I'm happy to see that the Chilean team of Letelier, Marín, and Mpodozis has begun this work in "Autopoietic and (M,R) Systems."

43. Rosen, *Life Itself,* 244.

44. Ibid., 246.

45. Varela, *Principles of Biological Autonomy.*

46. Rosen, *Life Itself* and *Essays on Life Itself,* 266–69.

47. Letelier, Marín, and Mpodozis, "Autopoietic and (M,R) Systems."

48. Varela, "Organism" and "Patterns of Life."

49. Weber and Varela, "Life after Kant."

50. Varela, "Organism" and "Patterns of Life."

51. Kant, *Critique of Judgement* (1951), §64–65.

52. Recently I rediscovered some e-mail correspondence Francisco and I had in June 1999 about this issue of teleology. It began because I pointed out to him that his commitment to phenomenology seemed inconsistent with his older position on teleology with Maturana. We had both independently been reading Kant and Jonas, and I asked Francisco whether he would still maintain his earlier anti-teleological stance in light of Jonas's argument that one cannot recognize something to be a living being unless one recognizes it as purposive and that one cannot recognize something as purposive unless one is an embodied agent who experiences purposiveness in one's own case. Francisco replied that he was "still quite suspicious" about this appeal to teleology, and hence to this way of linking phenomenology and biology, and that he preferred to "shift the accent" from teleology to original intentionality, understood as the sense-making capacity proper to autopoietic units. He saw this shift as a refinement of the "'Santiago school' move to introduce the equation life = cognition." It's clearly "silly," he said, to make cellular cognition just like animal cognition, but their "common root" is this basic sense-making capacity proper to autopoietic life. Appealing to sense-making, he suggested, was more "constructive" than appealing to the "elusive principle of purpose." Sense-making provides a strong link to intentionality, but "whether this turns into teleology," he said, "is another matter." This line of thought, however, struck me as unsatisfactory because "original intentionality" and "sense-making" are themselves arguably teleological notions. The issue is precisely how to analyze this teleology. So although the proposition "living is sense-making" may be an important elaboration of the equation life = cognition, it is insufficient to establish the anti-teleological stance with respect to Kant's and Jonas's notion of teleology. Six months later, in December 1999, in response to another e-mail of mine pressing him on this issue, Francisco indicated that as time had gone by, he had come to have a "broader view" and to see that "in a funny way you do recover a full fledged teleology . . . but this teleology is . . . intrinsic to life in action" and "does not require an extra transcendental source" in the Kantian sense. In other words, teleology, in the sense of self-organized natural purposiveness, can be seen as an empirical feature of the organism, based on its autonomy and sense-making, rather than only a form of judgment, as Kant had held. It's precisely this conception that Weber and Varela presented in "Life after Kant" and that they called *intrinsic teleology*.

53. See Varela, "Intimate Distances."

54. I borrow this phrase from Keller, *A Feeling for the Organism*.

55. Weber and Varela, "Life after Kant," 110.

56. Merleau-Ponty, *Phénoménologie de la perception*, 90; *Phenomenology of Perception*, 75.

57. Varela, "Neurophenomenology," 334.

Beyond Autopoiesis

Inflections of Emergence and Politics in Francisco Varela

JOHN PROTEVI

Francisco Varela's work is a monumental achievement in twentieth-century biological and biophilosophical thought. After his early collaboration in neo-cybernetics with Humberto Maturana (autopoiesis), Varela made fundamental contributions to immunology (network theory), artificial life (cellular automata), cognitive science (enaction), philosophy of mind (neurophenomenology), brain studies (the brainweb), and East-West dialogue (the Mind and Life conferences). In the course of his career, Varela influenced many important collaborators and interlocutors, formed a generation of excellent students, and touched the lives of many with the intensity of his mind, the sharpness of his wit, and the strength of his spirit. In this essay, I will trace some of the key turning points in his thought, with special focus on the concept of emergence, which was always central to his work, and on questions of politics, which operate at the margins of his thought. I will divide Varela's work into three periods—autopoiesis, enaction, and radical embodiment—each of which is marked by a guiding concept; a specific methodology; a research focus; an inflection in the notion of emergence; and a characteristic political question that specifies a scale of what I will call "political physiology"—that is, the formation of "bodies politic" at the civic, somatic, and "evental" scales. These terms refer to the formation of political states, politically constituted individuals, and their intersection in political encounters respectively.

The first period, marked by the concept of autopoiesis, runs from the early 1970s to the early 1980s and uses formal recursive mathematics to deal with synchronic emergence—that is, a focused behavior on the part of an organic system that is achieved via the constraint of the behavior of components of the system; synchronic emergence can be seen as the question of the relation of part and whole. The research focus is on identifying an essence of life. The political question here is the limit of using autopoiesis as a model for enacting social being. Varela sees autopoiesis as only an instance of a general mode of being, organizational closure; he restricts autopoiesis to cellular production—that is, to living systems bound by a physical membrane—and warns against using it as a model of social being. Here we see the question of the macro-scale of political physiology, the formation of a "body politic" in the classical sense, what we will call a "civic body politic." Varela refuses to countenance the use of autopoiesis as a model for social systems, I will argue, not so much for purely "cognitive" reasons, but because when autopoiesis is enacted, when it is the model for a way of social being, then social systems become obsessed with physical boundaries, leading to a fratricidal zero-sum competition. For him, systems above the cellular level—that is, neurological and immunological systems and social systems—are to be thought as informational systems with organizational closure. (Luhmann, however, will use the term "autopoietic" with regard to those systems as well.) The end result is that autopoietic enactment, in Varela's sense, is solely concerned with synchronic emergence (homeostatic part-whole relations) and is thereby unable to foster the condition for diachronic emergence in social and political dynamics (the emergence of novel patterns from the undoing of former patterns). I will argue that Varela implicitly holds that the historical changes and multiple causation of political systems must be thought in terms of a field whose dynamics are modeled with nonlinear differential equations, which is beyond the scope of autopoietic thought.

The second period, whose concept is that of enaction, spans the late 1980s and early 1990s and uses differential equations to model dynamic systems in order to deal with diachronic emergence, the production of novel functional structures. The research focus is embodied cognition. In this period we must distinguish two time scales of diachronic emergence: (a) the fast scale of the coming-into-being of a systematic focus of actual behavior from a repertoire of potential or virtual behaviors, and (b) the slow scale of the acquisition of the behavioral modules that form the virtual repertoire available to a system at any one time. The interplay of these scales requires that we think a "virtual self." The political question here is leisure: politics as the system controlling access to training for the acquisition of skills according to the differential access to leisure or free time. Here we see the meso-scale of political physiology, the formation of a somatic body politic as the resolution of the differential relations that structure a dynamic social-political-economic field, a process that is very crudely analogous to crystallization in a "metastable" supersaturated solution.

The third period, whose concept is that of radical embodiment, runs from the mid-1990s to Varela's premature death in 2001 and uses the methodology of neurophenomenology to discuss transversal emergence, the production of distributed and interwoven systems along brain-body-environment lines. The research focus is consciousness (both basic consciousness or "sentience" and higher-level reflective or self-consciousness) as it arises in the interaction of affect and cognition. With the turn to affect in theorizing concrete consciousness as enacted in distributed and interwoven brain-body-environment systems we approach the political questions of the other and concrete social perception and hence a micro-scale of political physiology, the formation of "evental" bodies politic or, perhaps less barbarically named, political encounters. As we will see, such encounters enfold all levels of political physiology, as a concrete encounter occurs in a short-term social context between embodied subjects formed by long-term social and developmental processes. More precisely—since "context" is too static—a political encounter, like all the emergent functional structures of political physiology, is the resolution of the differential relations of a dynamic field, in this case, one operating at multiple levels: civic, somatic, and evental. (Here we see the limits of the crystallization analogy: crystals form in homogeneous solutions, while political encounters coalesce in heterogeneous environments.)

Autopoiesis and Synchronic Emergence

Varela is perhaps best known for his early collaboration with Humberto Maturana in developing the concept of autopoiesis. This work, published in Spanish in 1973 and made known to the Anglophone community by a 1974 article and then by a 1980 monograph, is a classic of second-order cybernetics. In our terms, it is marked by a notion of "synchronic emergence," which is conducted in static part/whole terms. The concept of autopoiesis was developed to provide a horizon of unity for thinking living entities, rather than the haphazard empiricism of the "list of properties" model usually adopted (reproduction, metabolism, growth . . .). In other words, Maturana and Varela were trying to isolate an essence of life, an essence that would provide a viewpoint on life that is "history independent."[1]

To produce the concept of the essence of life, Varela and his colleagues distinguish organization (essence) and structure (historical accident). Organization is the set of all possible relationships of the autopoietic processes of an organism; it thus forms the autopoietic "space" of that organism.[2] Structure is that selection from the organizational set that is actually at work at any one moment.[3] Changes in the environment with which the system interacts are known as "perturbations" of the system. The system interacts only with those events with which it has an "interest" in interacting—that is, those events that are relevant to its continued maintenance of autopoietic organization (for example, nutrients). These events of interaction form a process of "structural coupling" that leads to structural changes in the system. These changes, as reactions to the perturbation, either reestablish the baseline state of the system (they reestablish the homeostasis of the system) or result in the destruction of the system qua living.[4] Homeostatic restoration thus results in conservation of autopoietic organization. From this essentialist viewpoint, the origin of life must be a leap into another register, a *metabasis eis allo genos*.[5] From the autopoietic perspective, questions of diachronic emergence have to be thought in terms of "natural drift," whose relation to autopoietic essential organization is problematic, as we will see. In any event, clearly autopoietic organization is synchronic emergence in which the whole arises from a "network of interactions of components."[6]

The difficulty here is that the assumption of organization as a fixed transcendental or essential identity horizon prevents us from thinking life as the virtual conditions for creative novelty or diachronic emergence. Life for autopoiesis is restricted to maintenance of homeostasis; creative evolutionary change is relegated to structural change under a horizon of conserved organization. If virtual organization is conserved for each organism, no matter the changes in its actual structure—one of the prime tenets of Maturana and Varela's autopoietic theory—then on an evolutionary time scale, all life has the same organization, which means all life belongs to the same class and has only different structure. As Katherine Hayles puts it, "Either organization is conserved and evolutionary change is effaced, or organization changes and autopoiesis is effaced."[7] Autopoietic theory gladly admits all this. "Reproduction and evolution do not enter into the characterization of the living organization"; evolution is the "production of a historical network in which the unities successively produced embody an invariant organization in a changing structure."[8] Although autopoietic theory, developed in the 1970s at the height of the molecular revolution in biology, performed an admirable service in reasserting the need to think at the level of the organism, it is clear that autopoiesis is locked into a framework that posits an identity horizon (organizational conservation) for (structural) change. To summarize: for autopoietic theory, living systems conserve their organization, which means their functioning always restores homeostasis; evolution is merely structural change against this identity horizon.

Let us focus on another key feature of autopoietic systems: the autonomy that they possess in virtue of their synchronic emergence. Their internal complexity is such that "coupling" with their environment or endogenous fluctuations of their states are only "triggers" of internally directed action. This means that only those external environmental differences capable of being sensed and made sense of by an autonomous system can be said to exist for that system, can be said to make up the world of that system. The positing of a causal relation between external and internal events is possible only from the perspective of an "observer," a system that itself must be capable of sensing and making sense of such events in *its* environment.

Quite soon after writing *Autopoiesis* with Maturana in 1973, Varela came to restrict the validity of the idea of autopoiesis to the cellular level, rejecting the use of autopoiesis as a concept for thinking social systems. In this period of his work, Varela distinguishes between autopoiesis, limited to physical production within the spatial border provided by a cellular membrane, and organizational closure, which can be applied to systems with an "informational" component. Varela thus comes to insist on the "complementarity" of two forms of explanation: autonomy versus control or, what amounts to the same distinction, autopoietic versus informational-symbolic explanations. In "On Being Autonomous: The Lessons of Natural History for Systems Theory" (1977), Varela insists that the autonomy and control perspectives are complementary. At this point, Varela is working with a recursion model of closure, where the "closure thesis" states that "every autonomous system is organizationally closed," and organizational closure entails the "indefinite recursion of component interaction."[9] Here Varela distinguishes cells as "physically independent units" from "systems where autonomy is expressed in an 'informational' way . . . [the] nervous and the immune system of animals, which are, as it were, cognitive systems in the macroscopic and microscopic domains of the organism."[10] It is this distinction between physical production enclosed in a physical space and the "information" of distributed systems that will lead him to restrict autopoiesis to the cellular level. "Information" of course must be in scare quotes, as the cognition Varela is talking about entails structural coupling and the triggering of autonomous response, rather than recovery of objective information.

Here Varela posits limits of "differentiable dynamic representation" (modeling of the changes in systems) due to the limited ability at the time to handle the differential equations necessary to model nonlinear dynamic systems and so opts for his self-referring, indefinite recursion model, which needs "an infinite-valued logic."[11] Operator trees are constructed and "circularity is captured through the solutions or eigenbehaviors of equations in this operator domain." This allows a "representation of autonomy which is not so abstract as indicational forms, and yet not so demanding of quantitative detail as in differentiable dynamic descriptions."[12] The paper closes with a clear statement of Varela's constructivism and anti-realism: "The contents of our reality are truly a reflection of the recursive biological and cognitive computations. . . . There is more a construction than a map."[13] We will see how what Varela would call the *autopoietic* enactment of this autonomous constructivism, whereby a system comes to focus on what it is already set up to see as being in its "interest" in maintaining its *physical* boundaries, will have disastrous effects in a time of civil war, when such an "epistemology" is instantiated in a political system producing mutually blind—and hence fratricidal—competing systems.[14]

In the meantime, we should stick with the question of the modeling of systems. In *Principles of Biological Autonomy*, Varela explains that he is attracted to dynamical systems models but finds them limited to the molecular level and suggests algebraic/formal recursion models as the most general kind to use in modeling larger systems. "The classical notion of stability in differentiable dynamics is the only well-understood and accepted way of representing autonomous properties of systems. . . . [We can find] excellent examples of the fertility of this approach for the case of molecular self-organization." However, this approach has a restricted validity: "An underlying assumption is, however, that there is a collection of interdependent variables, and it is the reciprocal interaction of these component variables that brings about the emergence of an autonomous unit. . . . [Thus] the differentiable dynamic description becomes a specific case of organizational closure." More precisely, the dynamic systems approach is of limited validity for organisms (and political systems, as we will see), where we find a number of interlocking and embedded informational or symbolic systems: "At the same time one finds the limitations imposed by [the differentiable framework]: More often than not, autonomous systems cannot be represented with differentiable dynamics, since the relevant processes are not amenable to that treatment. This is typical for informational processes of many different kinds, where an algebraic-algorithmic description has proven more adequate. Accordingly, the fertility of the differentiable representation of autonomy and organizational closure is mostly restricted to the molecular level of self-organization."[15]

The difference between the dynamical and the formal models depends on the difference between an abstract temporal approach and a concrete spatial approach. Varela refers to the dynamic approach of Eigen and Goodwin as one that focuses on a "network of reactions and their temporal invariances, but disregards on purpose the way in which these reactions do or do not constitute a unit in space." In this emphasis on physical boundaries and material production we see what leads Varela once again to insist on the need for complementarity between control and autonomy perspectives in which dissipative systems are treated as input-output fluxes. Although he claims there is some evidence that dynamic models are able to capture membrane formation, as in Zhabotinsky reactions, "it is still a matter of investigation how well the differentiable-dynamics approach can accommodate, in a useful way, the spatial *and* the dynamic view of a system." Beyond the molecular level we reach our cognitive limits, set by the state of knowledge at the time: "But it is in going beyond the molecular level, where we cannot rely on a strong physico-chemical background of knowledge, that the insufficiency of the differentiable framework appears, and thus the need to have a more explicit view of the autonomy/control complementarity, and an extension of differentiable descriptions to operational/algebraic ones."[16]

In other words, at the time of *Principles*, Varela thought that cellular autopoiesis could be thought dynamically and that while neurological and immunological processes were "borderline cases" higher level processes, organismic and social, could not be.[17] The key question is the ability to represent metastable (changeable, creative) systems. That is impossible in 1979 with the algebraic approach, and so we are left with a series of questions for further research:

> Clearly, both approaches cover somewhat *non*-overlapping aspects of systemic descriptions. Thus, it is necessary to have a way of dealing with plasticity and adaptation. Natural systems are under a constant barrage of perturbations, and they will undergo changes in their structure and eigenbehavior as a consequence of them. There is no obvious way of representing this fundamental time-dependent feature of system-environment interactions in the present algebraic framework. In contrast, the question of plasticity is a most natural one in differentiable frameworks because of the topological properties underlying this form of representation: hence the notions of homeorrhesis and structural stability in all their varieties. To what extent can the experience gained in the differentiable approach be generalized? How can notions such as self-organization and multilevel coordination be made more explicit in this context? Is category theory a more adequate language to ask these questions? These and many more are open questions.[18]

The political question in this first period is the extension of autopoiesis as a model for enacting social being, the question of the body politic in its classic sense, what we call the macro-scale of political physiology. Varela will reject all attempts at such an extension. The tension with Maturana on this point is evident in the 1980 English publication of *Autopoiesis*, where the authors note that they are unable to agree "on an answer to the question posed by the biological nature of human societies from the vantage point" of autopoiesis.[19] Varela's departure from Maturana is apparent in "On Being Autonomous," where autopoiesis is said to suggest a "universal feature" shared by many other types of systems—to wit, "organizational closure," which extends beyond physical systems to "informational" systems.[20] In "Describing the Logic of the Living," Varela is crystal clear: "Autopoiesis is a particular case of a larger class of organizations that can be called *organizationally* closed, that is, defined through indefinite recursion of component relations."[21] After insisting on some concrete sense of "production" to define autopoiesis, Varela drives home his point: "Frankly, I do not see how the definition of autopoiesis can be *directly* transposed to a variety of other situations, social systems for example."[22]

In a late interview, "Autopoïese et Émergence," Varela gives his reasons for resisting an extension of autopoiesis to the social:

> It's a question on which I have reflected for a long time and hesitated very much. But I have finally come to the conclusion that all extension of bio-logical models to the social level is to be avoided. I am absolutely against all extensions of autopoiesis, and also against the move to think society ac-cording to models of emergence, even though, in a certain sense, you're not wrong in thinking things like that, but it is an extremely delicate passage. I refuse to apply autopoiesis to the social plane. That might surprise you, but I do so for political reasons. History has shown that biological holism is very interesting and has produced great things, but it has always had its dark side, a black side, each time it's allowed itself to be applied to a social model. There's always slippages toward fascism, toward authoritarian impositions, eugenics, and so on.[23]

What is the key to the "extremely delicate passage" necessary to think social emergence while avoiding the "dark side" of the slide into fascism? First we should note the complete rejection of autopoietic social notions, while the notion of social emergence is less strongly condemned. I would argue that the difference lies in Varela's conception of autopoiesis as synchronically emergent, which locks out the sort of diachronic emergence we will study in the next section. If one could think the formation of civic bodies politic using dynamic systems modeling (something that for Varela at the time of *Principles* was con-sidered impossible, as we have seen), if one could see them as resolutions of the differential relations inherent in a dynamic field (again, something crudely analogous to crystallization in supersaturated solutions or lightning as the reso-lution of electric potential differences in clouds or weather systems as resolu-tion of temperature differences in air and water), then we would at least have the possibility of an "extremely delicate passage" in thinking political change. But without that possibility of novel production, modeled by dynamic systems means, autopoietic social systems, once formed and mature, construct a world only in their own image and, when locked in conflict with another such system, cannot ascend to an "observer" status that would see them both as parts of a larger social system. Instead, the two conflicting systems are locked in fratricidal combat, producing a torn civic body politic and in turn, civil war.

Let us turn here to "Reflections on the Chilean Civil War" for some historical detail about Varela's worries about the political misuse of "biological holism," or a misapplication of autopoiesis in enacting the macro-scale of political physiology, the formation of a society or body politic. In this discussion, "epistemology" is not a matter of neutral understanding but of enactment, of the bringing into being of a way of social living. The stakes are the highest possible for Varela in this deeply personal and emotional piece: "Epistemology does matter. As far as I'm concerned, that civil war was caused by a wrong epistemology. It cost my friends their lives, their torture, and the same for 80,000 or so people unknown to me."[24] Varela's analysis shows that Chile had become polarized into two separate worlds without communication—that is, one could claim, two "autopoietic" systems with no sense-making overlap, no means of mutual recognition, but only a concern with physical boundary maintenance: "The polarity created a continual exaggeration of the sense of boundary and territoriality: 'This is ours; get out of here.' "[25] I read this as Varela indicating the dangers of extending autopoietic notions to the social. The danger lies not in using autopoiesis as a means of understanding the social, but in using autopoiesis as a model in enacting a way of social being. An autopoietic social being is one focused on boundary maintenance, and this focus can create a fratricidal polarity.

The key to understanding Varela's prohibition on extending autopoiesis to social systems, that is, his move "beyond autopoiesis"—but not beyond neocybernetics as concern with organizational closure of informational systems—is to appreciate his warning against enacting the concern with physical boundary protection, which carries along with it the risk of falling into "polarization." Varela recounts his moment of insight when he overcame that polarization: "Polarity wasn't anymore this or that side, but something that we had collectively constructed"; political worlds, previously autonomous, had to be considered merely "fragments that constituted this whole." The problem, of course, is establishing the "observer" position that can use the notion of the interaction of organizationally closed informational systems to appreciate this larger whole encompassing the autonomous and mutually blind systems. Varela finds this position in Buddhist practice, with its necessity of stressing the "connection between the world view, political action and personal transformation." To avoid the fratricidal polarization of competing autonomous systems, relativistic fallibility is the key to the construction of a political world: "We must incorporate in the *enactment*, in the projecting out of our world views, at the same time the sense in which that projection is only one perspective, that it is a relative frame, that it must contain a way to undo itself."[26] Such flexibility, as we will see next, is available to a system producing a "virtual self" out of a multiplicity of coping resources, out of a repertoire of behaviors, but is foreclosed to the physical cellular systems to which Varela consigns autopoiesis. For that reason, the autopoietic model of cellular systems is disastrously misapplied when used to *enact* the macro-scale of political physiology, as in the brutally violent "epistemology" (*qua* way of social being) enacted by the conflicting sides in the Chilean civil war. To summarize Varela's position: *enacting* autopoiesis as a way of social being (as distinguished from using the concept of organizationally closed informational systems to *understand* a social situation) turns a social field into a polarized confrontation of systems seeking physical boundary maintenance; focused on synchronic emergence or part-whole relations, which it sees in zero-sum terms ("this is ours; get out of here"), such autopoietic enactment cannot foster the conditions for the diachronic emergence of historical novelty.

The Virtual Self and Diachronic Emergence

With this invocation of the key term "enactment," we can move to the second period of Varela's work, the late 1980s and early 1990s, in which the recursive models of systems Varela used in *Principles* under the acknowledged influence of Spencer-Brown's *Laws of Form* drop away as dynamical systems modeling makes progress, especially in connectionist work in cognitive science. Here we see that Varela's work develops a notion of "diachronic emergence" (emergence as the production of novel structures).[27] In this period, Varela broke into his own with a series of fundamental works on artificial life, immunology, and the status of the organism. This period culminates with his second most well-known work, *The Embodied Mind* (with Evan Thompson and Eleanor Rosch), the manifesto of the "enactive" school in cognitive science; this approach has been modified and developed in the work of, in particular, Thompson, Andy Clark, and Alva Noë.[28]

In this second period of his work, Varela deals with three "cognitive" registers: immunological, neurological, and organismic (which includes the previous two). We will concentrate on the intersection of the neurological and the organismic but should not forget Varela's groundbreaking work theorizing the immune system as a network; this work rejects the military metaphor of protection of interiority and resolves the paradoxes of self versus nonself rec-ognition that beset the classic concept.[29]

The inflection of emergence in the period of enactment or the virtual self is diachronic emergence, which operates at two temporal scales in both neurologi-cal and organismic registers. On the fast scale in the neurological register, we find resonant cell assemblies, which arise from chaotic firing patterns; on the fast scale in the organismic register, we see the arising of behavioral modules or "micro-identities" from a competition among competing modules. We can see that both these modes of diachronic emergence on the fast scale are resolutions of a dynamic, metastable differential field. While Varela concentrates on the fast scale, we will examine the slow scale, the acquisition of behavioral modules in those registers, for here we intersect the political question of leisure and access to training for acquisition of skills. The differential field here is the field of forma-tion of "somatic bodies politic," the meso-scale of "political physiology."

Working from connectionist models but rejecting their representationalist assumptions, Varela looks to resonant cell assemblies (RCA) as the neurological correlate for "micro-identities." The latter concept comes from phenomeno-logical reflection on the concrete life of the everyday. Following Heidegger and Merleau-Ponty in opposing a Cartesian heritage privileging self-conscious, reflective, and verbal reasoning as the essence of cognition, Varela will claim that most everyday life (of competent adults, to be sure) is accomplished in skilled, nonreflective comportments. Disruptive social encounters, however, lead to "breakdowns" in such everyday coping and can lead to reflective decision making or to the adoption of another skilled comportment.[30] The neurological correlates of breakdowns are a fall into a background of chaotic firing, out of which emerges a new RCA. This resolution of the differential field of widely distributed chaotic firing forms the basis for creativity in the arising within the organism of a triumphantly emergent comportment. There is no "choice" here, as the process of arising of an RCA is too fast for conscious reflection, which occurs in temporal chunks, so that RCA formation occurs "behind the back" of reflective consciousness. An RCA is the neurological correlate of what is described in other registers as a skill or agent or module, and the creative emergence occurs on the basis of the historical formation of a repertoire of behavioral modules.

We see here two important concepts: the virtual self and the enactment of world. As this repertoire is a distributed and modular system, at both the behav-ioral and the neurological levels, Varela will talk of a "virtual self" or "meshwork of selfless selves," as the subtitle of Varela's "Organism" puts it. The correlate of the virtual self, with its multiplicity of micro-identities, is the enacted world.

The laws of physics, or the regularities of the environment (the epistemological niceties that might distinguish these phrases need not concern us here), form only loose constraints for the worlds each organism brings forth or enacts in a process of "surplus signification." Here we see echoes of the sense-making at the heart of the autopoietic notion of "structural coupling," but with more ability to flesh out the neurological processes at work.

With these two concepts, as well as Heidegger and Merleau-Ponty, in mind, Varela and colleagues write in *The Embodied Mind*:

> The challenge posed by cognitive science to the Continental discussions, then, is to link the study of human experience as culturally embodied with the study of human cognition in neuroscience, linguistics, and cognitive psychology. In contrast, the challenge posed to cognitive science is to question one of the most entrenched assumptions of our scientific heritage—that the world is independent of the knower. If we are forced to admit that cognition cannot be properly understood without common sense, and that common sense is none other than our bodily and social history, then the inevitable conclusion is that knower and known, mind and world, stand in relation to each other through mutual specification or dependent coorigination.[31]

At this point I would like to shift from exposition to critical engagement by extending this series of challenges so that enaction is in turn challenged to examine the unconscious social grouping hiding in the "our" of "our bodily and social history." The challenge is to examine the historical and political system that distributes leisure and the access to training for learning of behavioral modules. A further challenge is to disabuse ourselves of the naive notion that all those modules are beneficial to the body that incorporates them, rather than beneficial to the power structure of the society. In other words, many people incorporate behavioral modules that hurt them, although they help reproduce inequitable social dynamics.[32]

We see the contours of this problematic in *Ethical Know-How*, published in the *Embodied Mind* period. The "constitution" of the "cognitive agent" is "a matter of commonsensical emergence of an appropriate stance from the entire history of the agent's life. . . . The key to autonomy is that a living system finds its way into the next moment by acting appropriately out of its own resources. And it is the breakdowns, the hinges that articulate microworlds, that are the source of the autonomous and creative side of living cognition."[33] Once again, we have to distinguish two temporal scales of diachronic emergence: "The moment of negotiation and emergence when one of the many potential microworlds takes the lead . . . the very moment of being-there when something concrete and specific shows up . . . within the gap during a breakdown there is a rich *dynamic* involving concurrent subidentities and agents."[34] This is the fast dynamic. If we are to engage Varela's work critically, we also need to thematize how the behavioral repertoire that provides the scope of those many potential microworlds has emerged over the slow scale of development, maturation, and learning. In other words, we must think the slow dynamic of structural coupling leading to the ontogenesis of the embodied subject; this process must be analyzed politically as the differential access to training. To bring out all its potential, Varela's insistence on autonomous organisms needs to be supplemented with an analysis, using social/political categories, of the distribution of access to training that allows differential installation of modules/agents/skills in a population of organisms.

The important thing is not to confuse autonomy and competence. A corporeal subject with a limited repertoire of capacities, or with a repertoire of disempowering habits, is still autonomous in the Varelean sense, as producing behaviors on the basis of environmental triggers or endogenous fluctuation. No matter how wide or narrow your repertoire of skills, no matter how powerful or weak you are in enacting them, you are no more autonomous than is any other organism in any one action. However, there is a difference in competence, how well your actions enhance your survival and flourishing and those of others, as well as a difference in the range of environmental differences you can engage and survive, thus preserving your autonomy for future encounters.[35] But you have to be trained to acquire many of these skills. As Varela puts it in *Ethical Know-How*: "The world we know is not pre-given; it is, rather, enacted through our history of structural coupling, and the temporal hinges that articulate enaction are rooted in the number of alternative microworlds that are activated in each situation."[36] Again, in order to develop Varela's insight more fully and thus to reach the full concrete reality of our social life, we have to analyze politically that history of structural coupling in terms of access to training to greater or lesser number and greater or lesser quality of skills opening microworlds.

The key to thematizing this meso-scale of political physiology is to think of downward causation in social emergence, the macro-scale of the body politic we referred to above. Picking up here on a contemporaneous essay written with Jean-Pierre Dupuy, Varela describes in *Ethical Know-How* the way upward causality allows for the emergence of social regularities: "Interactions with others. . . . Out of these articulations come the emergent properties of social life for which the selfless 'I' is the basic component. Thus whenever we find regularities such as laws or social roles and conceive them as externally given, we have succumbed to the fallacy of attributing substantial identity to what is really an emergent property of a complex, distributed process medi-ated by social interactions."[37] But here Varela is working with a formal model of synchronic emergence and has neglected the downward causality of these regularities, whether institutionalized in disciplinary intervention or distrib-uted as modulating "control," as they work in the slow temporal scale of the diachronic formation of somatic bodies politic in the context of a particular constellation of a civic body politic. As generations go by, we see a patterned differential social field, channeling perception, action, and affect along lines of social "roles." Varela has demonstrated only that laws, rules, institutions, etc. are emergently produced by upward causality in a synchronic emergence; he has neglected to show the downward causality effected by these regularities (which we could model by tracking the formation of attractors in a social space representing social "habits") and the way this socially enacted world structur-ally couples with, and guides, the ontogeny of the individual person. It's the pre-personal social field that needs to be thought. Persons are resolutions of the differential social field, concretions of the social field that form the affective topology of the person: the patterns, thresholds, and triggers of basic emotions or affective modules of fear, rage, joy, and so on as they interact with the cogni-tive topology of the person, the cognitive modules or basic coping behaviors that make up the everyday repertoire of the person.

Neurophenomenology and Transverse Emergence

In developing the practice of "neurophenomenology," a concept he coined in 1996, Varela begins his late period. It is in this period that the point of contact with politics appears in the question of concrete and affective social perception, the formation of the "evental" body politic or the political encounter, what we will call "transverse emergence." This latter term indicates the formation of a functional structure involving organic systems and environmental objects, including technological items, as we see in "extended cognition" involving the use of physical marks, ranging from simple scratches in clay tablets to calcula-tors, computers, and the like.[38]

In a late and very important article, collaborating for the last time with Evan Thompson, Varela writes: "Neural, somatic and environmental elements are likely to interact to produce (via emergence as upward causation) global organism-environment processes, which in turn affect (via downward causa-tion) their constituent elements."[39] There is a slight terminological nuance here, as Varela has always distinguished "environment" (as objectivist or realist) from "world" (as enactive). We are to read this distinction as maintaining that the "environment" (= "laws of nature" or physical regularities) provides constraints on world-making: it constrains but does not optimally specify the enaction of worlds. Thus, to invoke the classic example from *The Embodied Mind*, light obeys laws of physics, but that only provides constraints on the construction of many different enacted color-worlds, which track lines of natural drift. The precision introduced in this article is that we do not see structural coupling between organism and world but between organism and environment, with the latter coupling being the process of the enactment of world. With this in mind, we note that Thompson and Varela specify three dimensions of "radical embodiment."

1. Organismic regulation in which affect appears as a "dimension of organismic regulation, . . . the feeling of being alive . . . inescapable affective background of every conscious state."

2. Sensorimotor coupling, where "transient neural assemblies mediate the coordination of sensory and motor surfaces, and sensorimotor coupling with the environment constrains and modulates this neural dynamics. It is this cycle that enables the organism to be a situated agent." Insofar as "situated agent" means "that which enacts a world," we see that coupling with the environment constrains and enables world-making.

3. Intersubjective interaction, whereby "the signaling of affective state and sensorimotor coupling play a huge role in social cognition. . . . Higher primates excel at interpreting others as psychological subjects on the basis of their bodily presence (facial expressions, posture, vocalizations, etc). . . . Intersubjectivity involves distinct forms of sensorimotor coupling, as seen in the so-called 'mirror neurons' discovered in area F5 of the premotor cortex in monkeys. . . . There is evidence for a mirror-neuron system for gesture recognition in humans, and it has been proposed that this system might be part of the neural basis for the development of language."[40]

We should note here that the thought of intersubjectivity in Varela's late period stems from the notion of "the other" as developed in the theory of the recognition of the alter ego, based on Husserl's Fifth Cartesian Meditation (although supplemented by the recognition of recent research into mirror neurons). For example, Varela writes in a popularizing article from 1999:

It is best to focus on the *bodily* correlates of affect, which appear . . . as directly felt, as part of our *lived body*. . . . This trait . . . plays a decisive role in the manner in which I apprehend the other, not as a thing but as another subjectivity similar to mine, as alter ego. It is through his/her body that I am linked to the other, first as an organism similar to mine, but also perceived as an embodied presence, site and means of an experiential field. This double dimension of the body (organic/lived; *Körper/Leib*) is part and parcel of empathy, the royal means of access to social conscious life, beyond the simple interaction, as fundamental intersubjectivity.[41]

To see how the problematic of the "other" is an abstract "philosopher's problem," let us note that in *The Embodied Mind* Varela and his collaborators, Evan Thompson and Eleanor Rosch, cite Rosch's research into categorization, where, in a 1978 article, she posed a "basic level" of perception/action/linguistic naming in a hierarchy of abstraction. This basic level is, in her example, "chair" rather than "furniture" or "Queen Anne." In the same article Rosch proposed a "prototype" theory for internal category structure—rather than an ideal exemplar, we have concrete prototypes by which we judge category membership by how close or far an object is to our prototype, not whether or not it satisfies a list of necessary and sufficient conditions that we carry around with us. If we adopt Rosch's model, in concrete social perception we are never faced with the Husserlian problem, "Is this just a thing or is it an alter ego?" which we resolve by distinguishing between things and subjects. Rather we are always confronted with other people at "basic level" social categories appropriate to our culture: for us today, the famous age, size, gender, race, and class system. So we never see another "subject"; instead, we see over here, a middle-aged, small, neat, fit, professional black woman (Condoleeza Rice, let's say), or an elderly, patrician, tall white man (George H. W. Bush, let's say).

So we have to say that Varela's discussion in "Steps" is abstract, which is revealed by his use of "his/her." In our society, we never *perceive* a "subject" we can call "his/her"; we can posit such a creature, but that's a refined political act of overcoming our immediate categorization process, by which we perceive gendered subjects, to construct an abstraction we can call a nongendered "intersubjective community" or "humanity" or some such. While this might be a worthy ethical ideal for which we can strive, it's not what we perceive "at first glance."

It's not that we are completely without guidance here regarding social perception. In "At the Source of Time," Varela and Depraz mention what would need to be fleshed out: among the five components of affect the first is "a precipitating *event*, or trigger that can be perceptual (a social event, threat, or affective expression of another in social context) or imaginary (a thought, memory, fantasy . . .) or both."[42] In other words, the social trigger has to be recognizable, based on the ontogeny of the perceiving subject. As we claim above, this ontogeny has to be thought as a resolution of a pre-personal dynamic differential social field. After learning our mid-level social categories,[43] we never immediately encounter an "other," only concrete people we locate in complex social categorization schemes. The encounter with the "other" is the result of an abstraction, a working up of the initial encounter, abstracting away from the "mid-level" categories of concrete social perception.[44]

Let us conclude by returning to "Reflections on the Chilean Civil War," where Varela provides an example of mid-level categories in concrete social perception and affect: "I remember very well that the soldier, whom I saw ma-chinegunning the other fellow who was running down the street, was probably a 19-year-old boy from somewhere in the South. A typical face of the people of the South. . . . I could see in his face what I had never seen, a strange combina-tion of fear and power."[45]

Varela's reminiscence rings true to concrete social perception. He didn't see a neutral "subject," an "other"; he saw a Southern Chilean boy of nineteen, a concrete person who is gendered, aged, and racially or at least ethnically marked. In that marking, and in the perception of a new affective state on the soldier boy's face, that "strange combination of fear and power," we engage all scales of political physiology: the macro-scale of a civic body politic torn apart in civil war; the meso-scale of the development of the repertoire of behavioral modules, as the boy is marked by this affective combination; and the micro-scale of politi-cal encounter, mediated by affect and cognition on Varela's part as this assem-blage or momentary transversal emergence arises: street, gun, soldier, shooting, running, dying, observing. Our challenge is to negotiate the "extremely delicate passage" of social emergence that would let us think through the interchanges among all levels of political physiology in this haunting scene—civic, somatic, and evental—at once.

Notes

1. Varela, Maturana, and Uribe, "Autopoiesis," 187.
2. Maturana and Varela, *Autopoiesis and Cognition*, 88; scare quotes in original.
3. Ibid., xx, 77, 137–38. See also Hayles, *How We Became Posthuman*, 138, and Rudrauf et al., "From Autopoiesis to Neurophenomenology," 31.
4. Maturana and Varela, *Autopoiesis and Cognition*, 81.
5. "The establishment of an autopoietic system cannot be a gradual process; either a system is an autopoietic system or it is not" (Maturana and Varela, *Autopoiesis and Cognition*, 94).

6. Varela, Maturana, and Uribe, "Autopoiesis," 187. In "Not One, Not Two," Varela also notes the synchronic emergence of wholes from the interaction of parts (63).

7. Hayles, *How We Became Posthuman*, 152.

8. Maturana and Varela, *Autopoiesis and Cognition*, 96, 104.

9. Varela, "On Being Autonomous," 79.

10. Ibid.

11. Ibid., 81.

12. Ibid., 82.

13. Ibid.

14. I am not claiming that *all* systems of organization closure—for instance, those that are in Varela's terms *informational*—are dangerous social models. Thus I am not arguing that Varela's warning applies generally to social systems, which Luhmann treats as "autopoietic" by redefining them in terms of self-referential communicative events. See Luhmann in this volume.

15. Varela, *Principles of Biological Autonomy*, 203.

16. Ibid., 204–5. Emphasis in original.

17. Ibid., 205. Advances in computer simulation will later allow for the dynamic modeling of cooperation and competition in the formation of resonant cell assemblies, as we can see by 1991 in Varela, Thompson, and Rosch, *The Embodied Mind*.

18. Varela, *Principles of Biological* Autonomy, 205–6. Emphasis in original. Robert Rosen takes up category theory (cf. *Life Itself*); Varela drops the formalization as more adequate dynamical models appear.

19. Maturana and Varela, *Autopoiesis and Cognition*, 118.

20. Varela, "On Being Autonomous," 79.

21. Varela, "Describing the Logic of the Living," 37; emphasis in original.

22. Ibid., 38; emphasis in original.

23. Varela, "Autopoïese et émergence"; my translation.

24. Varela, "Reflections on the Chilean Civil War," 19.

25. Ibid., 16.

26. Ibid., 18–19; emphasis added.

27. In a dialogue with Cornelius Castoriadis, Varela specifies that such emergence is neither aleatory nor calculable (Castoriadis, *Post-scriptum sur l'insignifiance*, 113).

28. Thompson, *Mind in Life*; Clark, *Being There*; Noë, *Action in Perception*.

29. See, for example, Varela and Coutinho, "Immunoknowledge."

30. Varela, "Organism" and "Making It Concrete."

31. Varela, Thompson, and Rosch, *The Embodied Mind*, 150.

32. Young's *Throwing like a Girl* is a classic critique of the privileged and empowered masculine corporeal subject presupposed in Merleau-Ponty's analyses. Young shows how many feminized corporeal subjects experience parts of the world as anxiety-producing obstacles, the "same" parts that a competent masculinized subject will encounter as amusing occasions for the demonstration of competence.

33. Varela, "Making It Concrete," 11.

34. Ibid., 49.

35. Students of philosophy might wish at this point to take up the connection to Spinoza: "What can a body do? How can it be affected?"

36. Varela, *Ethical Know-How*, 17.

37. Dupuy and Varela, *Understanding Origins*, 62.

38. Regarding the emergent functions in the interplay of animal physiology and the structures they build, see Turner, *The Extended Organism*. Regarding human-technology interfaces, see Clark, *Natural Born Cyborgs*, and Hansen, *Bodies in Code*.

39. Thompson and Varela, "Radical Embodiment," 424.

40. Ibid.

41. Varela, "Steps to a Science of Inter-Being," 81, original emphasis; see also Varela and Depraz, "At the Source of Time," where we read of "a primordial duality, a rough topology of *self-other*." Original emphasis.

42. Varela and Depraz, "At the Source of Time." Original emphasis.

43. Rodgers and Hammerstein in *South Pacific*: "You've got to be taught . . . carefully taught!"

44. This is where we need more empirical work on humans and mirror neurons. With monkeys we know that it is simply intra-specific. I want to ask if that's the case for humans or if our historical-cultural bodily formation (what I'm calling "political physiology") doesn't set even our mirror neuron empathizing at socially constructed "mid-level" categories. Do we "dehumanize" enemies in warfare (we can kill them because they're inhuman vermin, insects, rats, etc.), or is "simple" racialization enough? Is it that the inferior races are humanly liminal, at the border of animals? For more on mirror neurons, see Brigham in this volume.

45. Varela, "Reflections on the Chilean Civil War," 18.

System-Environment Hybrids

MARK B. N. HANSEN

Instead of searching for mechanisms in the environment that turn organisms into trivial machines, we have to find the mechanisms within the organisms that enable them to turn their environment into a trivial machine.

—HEINZ VON FOERSTER

It seems to me . . . that autopoiesis deserves to be rethought in relation to entities that are evolutive and collective, and that sustain diverse kinds of relations of alterity, rather than being implacably closed in upon themselves. Thus institutions, like technical machines, which, in appearance, depend on allopoiesis, become ipso facto autopoietic when they are seen in the framework of machinic orderings that they constitute along with human beings. We can thus envision autopoiesis under the heading of an ontogenesis and phylogenesis specific to a mechanosphere than superimposes itself on the biosphere.

—FÉLIX GUATTARI

Instead of the *surfaces* so typical of first modernities—the "domains" of science, of economy, of society, the "spheres" of politics, values, norms, the "fields" of symbolic capital, the separate and interconnected "systems" so familiar to readers of Luhmann, where homogeneity and control could be calmly considered—we are now faced with the rather horrible melting pots so vividly described by historians and sociologists of science.

—BRUNO LATOUR

Picking up on Latour's remark concerning the "rather horrible melting pots" we now face in our highly technologized, "posthumanist" world, I offer this meditation as an appeal for a flexible and adaptive understanding of what the legacy of neocybernetics might be for those of us seeking to make sense of questions of agency, identity, complexity, and evolution in the face of the massive technical distribution that currently characterizes our "cognition in the wild."[1] The straightforwardly historicist assumption that orients my thinking here is that the complexity of the world has undergone a double transformation sometime in the fairly recent past: first, worldly (environmental) complexity has become so intense and so messy (hence Latour's invocation of "horrible melting pots") that any effort to reduce it through selection by systems (or their avatars) cannot ignore the agency that is wielded by the environment; and second, the operation of this environmental agency is now predominantly and ever increasingly technical, meaning that system function is irrevocably permeated by technicity from the environment.

As I see it, this technical intensification of complexity and the irrevocable agency it accords the (to some extent unreduced) environment calls on the contemporary theorist to abandon the desire for purity that has informed neocybernetics from Heinz von Foerster's cybernetics of cybernetics to Luhmannian systems theory and beyond. Contemporary neocybernetic thinking must adapt the significant resources of neocybernetics to address a changed world (or environment) that quite simply *resists* the reduction of complexity so central to both von Foerster's and Luhmann's projects. This injunction, I want to emphasize, does not in any way entail abandoning what we in this volume collectively take to be the central motif of neocybernetics—namely, the operational closure of self-referential systems; indeed, as I see it, the increased complexification of the environment makes operational closure, no matter how provisional and heterogeneous it may now in actuality be, all the more urgent.

What the injunction to adapt *does*, however, call on us to do, is to move—*in concert with the major players in neocybernetic theory*—beyond the polarization of open and closed systems and focus instead on the negotiation of multiple, diverse boundaries made necessary by the hypercomplexification of the environment. By italicizing the above lines, I mean to emphasize that my call for flexibility is perfectly in accord with the spirit, if not always with the letter, of the work of major neocyberticians, including von Foerster and above all Luhmann.

Indeed, Luhmann stresses his own break with older forms of systems theory on precisely the topic of open and closed systems. "Newer developments in systems theory," he pronounces in *Social Systems*, "no longer interpret the distinction between open and closed systems as an opposition of types but rather regard it as a relationship of intensification. Using boundaries, systems can open and close at the same time, separating internal interdependencies from system/environment interdependencies and relating both to each other. Boundaries are thus an evolutionary achievement par excellence; the development of all higher-level systems, above all the development of systems with internally closed self-reference, presuppose them."[2] By emphasizing the contingency of any system's selectivity, Luhmann shows how the same environmental factors can support the constitution of different boundaries and thus give rise to different systems. From here, I would suggest that it need only be a small step to the recognition that, alongside systems proper (those self-referential processes that can maintain total closure to environmental agency and can reproduce their constitutive elements entirely through their own operation), there exist "system-environment hybrids" (SEHs) that realize their autonomy at a higher level of inclusiveness—which is to say *through a constitutive relation with alterity*—and that, for this reason, cannot be qualified as autopoietic. One key question that arises from such a claim—a question that clearly divides the authors of this volume (along with the main players in the articulation of neocybernetics)—is to what extent the environment, and specifically the operation of SEHs, interferes with the smooth functioning of system autopoiesis. As we shall see, much depends on how one understands the balance of power between systems and SEHs in our contemporary technosphere.

For my part, I shall try in what follows to combine a defense of the motif of operational closure, modified in certain respects from its canonical articulation by Maturana and Varela, with an effort to account for the undeniable agency of the environment as to some extent a nonreducible force of alterity. My claim is that even if this force of alterity may not directly enter into the operation of (some) systems (to the extent that such operation can be isolated from their worldly interactions, which itself remains an open question), it certainly does inform their cognitive operations in the world. In this sense, when (the majority of) systems operate in any concrete context, they *always* and *necessarily* do so in conjunction with a technical environment whose agency cannot be reduced to mere perturbations—whose agency not only acts in ways other than to maintain system reproduction but also more generally remains beyond the scope and mastery of the systemic perspective.

Let me say something more specific concerning why closure remains such a crucial motif, particularly in light of the hypercomplexification of the contemporary technosphere. In concert with Luhmann, I believe that selection is key to instituting difference into what would otherwise remain undifferentiated chaos: "One must distinguish the incomprehensible complexity in a system (or its environment) that would result if one connected everything with everything else, from determinately structured complexity, which can only be selected contingently. And one must distinguish environmental complexity . . . from system complexity . . . ; the system complexity is always lesser and must compensate by exploiting its contingency, that is, by its pattern of selections. . . . The *difference* between two complexities is the real principle compelling (and therefore giving form to) selection."[3] Although I concur with the main claim here regarding contingently selected complexity (which I take to be an alternate statement of the *key* principle of neocybernetics), I remain less convinced that contingent selection need take the form of a discrete distinction between system (agent of selection) and environment (undifferentiated ground of selectivity) and also that this yields a "completely different understanding of complexity" antithetical to the treatment it receives in contemporary theories of emergence. Not only does this argument strike me as circular and in an odd sense instrumental (since the difference produced by contingent selection becomes the motivation for that very selection), but it also, and of more immediate consequence, has the effect of rendering the very structure of the neocybernetic edifice remarkably defensive in nature: hobbled by its deficit of complexity vis-à-vis the environment, the system must unceasingly struggle to defend its fragile (and, as I shall argue, in some sense, false) autonomy. As any reader of Luhmann's systematic statements cannot but admit, the motif of system preservation (autopoiesis) takes precedence over any interest in accounting for what's actually going on in the environmental "wild." As I see it, then, the bottom line is that this fundamentally defensive positioning makes an unfortunate starting point from which to grapple with the undeniable (technical) complexification, which is to say with the unprecedented material agency, of the world.

In this respect, my heart is with theorists like Katherine Hayles, Félix Guattari, and Andy Clark, who, in various ways, have sought to move beyond the motif of autopoietic closure. For all of these theorists, autopoietic closure remains insufficiently dynamic to grapple with the basic operation of change that is everywhere at work in the world. Thus, Hayles has recently called for a rethinking of the permeability of boundaries of all kinds; Guattari has urged a shift from autopoiesis to machinic heterogenesis; and Clark has promoted a multiplication and complexification of boundaries in the place of their defensive ossification. Yet while I concur substantially with many of the claims offered by these theorists, I remain concerned that the effort to loosen the grip of autopoietic closure risks losing the motif of closure *tout court*. To my mind, some conception of closure, however provisional and non-autopoietic it may turn out to be, is absolutely necessary to introduce differentiation into the undifferentiated flows of the contemporary technosphere. It is in this respect that I consider myself a proponent of neocybernetics and an advocate of the irreducibility of selectivity (not to mention a fellow traveler with my colleagues in this volume): through its minimal insistence on the motif of closure (which must, as I shall argue shortly, be sharply distinguished from *autopoietic* closure), neocybernetic thinking can furnish the necessary conceptual arsenal to specify the heterogeneous correlations between humans and technics in our hypercomplex contemporary world. As I see it, neocybernetic thinking allows a more fine-grained and more rigorous analysis than Hayles, Guattari, or Clark of the concrete ways in which boundaries are not only multiple and complex, but also permeable or impermeable *depending on the scale being explored.* In sum, then, without necessarily concurring with every element of neocybernetic discourse as it has been developed by its numerous theoreticians, I shall make the case that the neocybernetic principle of closure is particularly crucial for us today as we seek to negotiate the new SEHs in and through which we live. For me, in short, closure becomes all the more imperative (if all the more provisional and scale-specific) as the technical complexity of the environment increases.

What is at issue in contemporary environmental complexification is the technical distribution of cognition that has revolutionized not simply the various cognitive sciences but also first and foremost the actual experiential domains they study. In today's complex computational world, countless instances of human agency—even those as mundane as making online credit card and mortgage payments, monitoring information about the weather or the stock market, even writing letters and sending messages—occur against the backdrop of complex computational infrastructures, which geographer Nigel Thrift has christened with the felicitous name of the "technological unconscious."[4] Such instances witness the operation of hybrid agencies, which, when viewed from the perspective of the human agent or agents involved, comprise system-environment hybrids. I submit that this human-anchored, even human-centered (though not necessarily humanist) perspective remains the predominant one for the vast majority of hybrid, technically distributed forms of agency: most instances of technical distribution of cognition, that is, involve the use of technics to expand the agency *of human mindbodies*, whether individual or collective in scale.

This state of affairs—the proliferation of SEHs where the system in question *is the human mindbody*—presents a unique problem for contemporary cultural theorists: how can one recognize the certain consistency, perhaps even the autonomy, of the (individual or collective) human mindbody and at the same time account for the certain non-autonomy that accrues from its unavoidable reliance on the agency of informationally complex environments to achieve its cognitive tasks?

The answer I would propose involves two revisions of the neocybernetic motif of autonomy:

(1) Our conception of autonomy must be multiplied and differentiated so that different forms of closure apply to different levels of cognitive operation. This necessarily involves a decoupling of closure from autopoiesis or autopoietic closure, along the lines proposed by Varela in *Principles of Biological Autonomy.*

(It also entails a break with Luhmann's effort to preserve the pattern of autopoiesis across all systems, meaning-producing as well as non-meaning-producing.)[5] Thus, while the closure of parts of the human agent may remain autopoietic, this latter's operation—or better, its co-operation—in any concrete cognitive situation will involve not just this autopoietic closure, but also various other, differently scaled, and more provisional non-autopoietic closures, pertaining to the higher order and psychic functions of the human being, as well as the transindividual co-operation with the "technological unconscious." As we will see, the flexibility of our biological makeup, of the organismic closure that gives us our embodiment, is crucial to our capacity to combine multiple and heterogeneous closures in order to act in the world.

(2) Our picture of both system operation and SEH operation must be made far more complex to do justice to this multiplication and differentiation of closure. In order to analyze adequately the cognitive operation of any system or SEH today, we need to take account of *all* of the heterogeneous closures that combine to effectuate its agency in any concrete situation. In the case of systems, this may involve taking account of the autopoietic closure of human mindbodies (which Luhmann simply assimilates to the organism and assumes as part of the environment of psychic systems) alongside the non-autopoietic closure of a meaning system like language or print textuality. In the case of SEHs, which, I want to reiterate, predominate in today's complex world, an adequate account would involve yet more levels of closure: not simply the closures just mentioned, but also the hybrid, provisional closure effectuated by (say) the use of the computer to perform what Hayles has long referred to as cognitively sophisticated tasks. In this last instance, we can see how far closure has come from both autopoiesis and autonomy, at least as it has been traditionally treated: more or less in line with Luhmann's minimal definition, closure here simply indicates that the operation of an SEH does not involve all aspects and possibilities of a system or of each system involved and that it does not involve all aspects of the environment, all possibilities for environmental agency. Because SEHs are produced by selections (although, it is important to note, selections that take place *across* the system-environment divide), they can and must be considered to be closed, with closure here meaning simply that the cognitive operation they perform is due to the various selective agencies (systemic *and* environmental) they bring—provisionally and situationally—together. What is at issue in such cases is a set of functional closures, rather than some overarching instance of autopoietic closure.

To rethink closure across the system-environment divide, as I propose here, requires a reformulation of the specificity of systems and their distinction from what lies beyond their boundaries. Thus, for me, the difference between a system and a system-environment hybrid does not involve autopoietic closure so much as the degree of mastery possessed by the system: whereas systems in the narrow sense enjoy cognitive mastery over the aspects of the environment to which they are coupled or interactionally open, this condition no longer holds in the case of SEHS. In reference to systems, we can say that the environment has trivial agency or, perhaps more accurately, that the environment functions (or is attended to) predominantly if not exclusively as a *support* for system agency; in the case of SEHS, by contrast, we cannot avoid recognizing the nontrivial nature of the environment's agency. Its role no longer being simply to provide support for system reproduction, the environment in SEHS has real agency, defined as cognitive operationality beyond the maintenance of system identities or autopoiesis.

Although this reformulation of the neocybernetic principle of closure admittedly wreaks havoc on the neat Luhmannian distinctions that it adapts, the resulting messiness is simply the cognitive cost of technico-material complexification. (Recall the "rather horrible melting pots" invoked by Latour!) In what follows, I shall first seek to clarify this messy picture by considering some recent accounts of cognitive distribution. Then, I shall construct an alternate neocybernetic genealogy, set in motion by Varela's decoupling of closure from autopoiesis and extended in crucial ways by the work of political philosopher and psychoanalyst Cornelius Castoriadis and the French philosopher of technology and biophenomenologist Gilbert Simondon. The advantage of this genealogy is that it manages to toe the line between the hard program of neocybernetics (exemplified by Luhmann) and the call for a dissolution of boundaries (exemplified by Hayles), between the necessity of closure and the recognition that technico-material complexification resists reduction. Introduced not as a specific program for contemporary cultural theorists but rather as an appeal for flexibility in our assumption of the neocybernetic legacy, this alternate genealogical lineage instances one (hopefully) successful combination of a (transformed) conception of closure with a commitment to the nontrivial agency of the environment.

Consider the reversal of the arrow of complexification telescoped in the passage from Heinz von Foerster given as an epigraph above. Citing it in his introduction to von Foerster's *Observing Systems*, Varela indicates that in its context within von Foerster's 1970 essay "Molecular Ethology," the remark is intended to describe what occurs in "a typical conditioning experiment" of the sort that behavioral psychology was carrying out in the 1960s.[6] Accordingly, von Foerster's claim serves to indicate just how central the line of questioning pursued here has *always been* for neocybernetic thinking. Far from providing a succinct expression of the methodological orientation of neocybernetics toward decomplexification of the environment, von Foerster's remark begs the question whether the environments of living systems can ever be rendered "trivial."[7] As I see it, the factors motivating von Foerster's skepticism concerning reduction have only grown more powerful with time. In the face of the hyperaccelerated technological change underwriting the "disorientation" that now characterizes our phenomenological interaction with the environment (as critics like Paul Virilio and Bernard Stiegler have underscored), it seems downright futile even to proffer the possibility of a trivialization of the environment. Moreover, the very picture of the system exerting complete control over its own self-restructuring in response to environmental perturbations seems highly idealistic in the context of the massive infrastructural role played by technologies in our world today. If our agency depends increasingly on the "technological unconscious" (Thrift), surely some dimension of it must involve the nontrivial operation of environmental processes that are simply beyond our control.

It is precisely this intuition that informs Katherine Hayles's skepticism toward second-order cybernetics. If, as Hayles suggests, the neocybernetic picture of recursivity remains locked in a vicious circle, unable to open out into a spiral that would yield "dynamic hierarchies of emergent behaviors," that is precisely because of its methodological postulate of closure: so long as recursivity does "no more than turn back on itself to create autopoietic systems that continually produce and reproduce their organization," it cannot, apparently for reasons of principle, yield anything new. That is why Hayles understands the significance of today's technical hyperacceleration in terms of its challenge to boundaries of all kinds, and especially to the boundary of the skin: "Boundaries of all kinds have become permeable to the supposed other. Code permeates language and is permeated by it; electronic text permeates print; computational processes permeate biological organisms; intelligent machines permeate flesh. Rather than attempt to police these boundaries, we should strive to understand the materially specific ways in which flows across borders create complex dynamics of intermediation."[8]

Before simply acceding to this sweeping claim, we would do well to pinpoint exactly what is at stake in the boundary violations inventoried by Hayles. For when code permeates language (as it does in the computational hosting of writing in a natural language, such as the writing you are now reading) or when electronic text permeates print (as it does in various hypertext hostings of specific print texts or conventions), what is at issue is not a wholesale fusion of the two forms of materiality, nor a total de-differentiation of the form of natural language or print text, but a certain supplementation and a certain provocation that may change the functionality of language or print but not its systematicity. What I mean by this is that the system of a natural language like English is in no way *operationally modified* by the intrusion of elements of code, just as the conventions of print textuality are not somehow de-specified by the embedding of print within an electronic environment. Rather than a wholesale leveling of distinctions between system and environment, what is thus needed is a more nuanced account of the way that the boundary crossings Hayles pinpoints—and the forms of intermediation they yield—modify (almost always by expanding) the function of a system without necessarily transforming its constitutive organization.

Such an account becomes even more urgent when we turn to the domain of the living and ask what exactly is involved when computational processes permeate biological organisms, when intelligent machines permeate flesh, as they certainly do, in manifold and well-nigh unrepresentable ways, in our world today. In his recent book, *Natural Born Cyborgs*, cognitive scientist Andy Clark turns his attention to this very question, sketching an answer that, however much it may resonate with Hayles's call for a farewell to neocybernetics, not only downplays the novelty of human-machine mergers (in concert with Stiegler and Manuel DeLanda, Clark insists on the fact that the living human being has always been dependent on cognitive scaffolding), but also—and of more consequence for us here—preserves the specificity of system operation across the apparently posthuman mergers that are occurring everywhere today. Beyond suggesting that we have always (or have never) been posthuman, Clark's position approaches the question of boundary violation less through specific contemporary technologies (and, his title notwithstanding, the concept of the "cyborg") than through the biological flexibility that characterizes (and has always characterized) the living, embodied human being. We are, Clark intones repeatedly throughout his book, "creatures whose minds are special precisely because they are tailor-made for multiple mergers and coalitions."[9]

Clark emphasizes that our special "ability to enter into deep and complex relationships with nonbiological constructs, props, and aids" is more than a mere technical supplementation and that it can be traced back to our biological flexibility: "This ability," Clark specifies, "does not depend on physical wire-and-implant mergers, so much as on our openness to information-processing mergers. Such mergers may be consummated without the intrusion of silicon and wire into flesh and blood, as anyone who has felt himself thinking *via* the act of writing already knows. The familiar theme of 'man the toolmaker' is thus taken one crucial step farther. Many of our tools are not just external props and aids, but . . . are deep and integral parts of the problem-solving systems we now identify as human intelligence."[10] While hard-line neocyberneticians will bridle at Clark's mention of information-processing mergers, let us be clear about what Clark has in mind here: in the kinds of cognitive distribution he seems to be envisioning, it is the *flexibility* of the human mindbody that allows us to modify our own cognitive operations through our co-functioning with external, technical scaffolds such as the computer-cum-word-processor. Far from being penetrated by information from the outside, we have the capacity to use technologies to expand the sphere of our own "in-built" cognitive capacities.

I want to stress that Clark's emphasis falls squarely on the flexibility *of the human mindbody* and not on the agency of the environment. His aim is to describe the mindbody in the age of its technical expansion, rather than to explore the direct agency of today's complex computational environments. For Clark, then, the extent of the environment's role is to furnish an impetus impelling us to rewire our brains—that is, to exercise a capacity that originates in and belongs squarely to our biological makeup:

> Our natural proclivity for tool-based extension, and profound and repeated self-transformation . . . explains how we humans can be *so very special* while at the same time being not so very different, biologically-speaking, from the other animals with whom we share both the planet and most of our genes. What makes us distinctively human is our capacity *to continually restructure and rebuild our own mental circuitry,* courtesy of an empowering web of culture, education, technology, and artifacts. *Minds like ours are complex, messy, contested, permeable, and constantly up for grabs.*[11]

For Clark, our contact with informationally rich environments is in itself sufficient to spur the restructuring of our brains; no information need be transferred. Yet, by way of contrast to Luhmann (with whom Clark's emphasis would otherwise seem of a piece), the important point here is that this restructuring can be triggered by the environment *only because of the flexibility of our brains,* which is to say only because of the margin of indetermination that is built into our operationally closed biological organization. The role of this flexibility cannot be stressed enough. For if "cultural scaffolding [can] change the dynamics of the cognitive system in a way that opens up new cognitive possibilities,"[12] this takes place *without* any alteration of the biological makeup of the human being, without any informational intrusion into our constitutive operational closure. Cultural scaffolding modifies the human cognizer without impacting its biological closure.

How would this picture change if we were instead to place our emphasis on the agency of the environment itself? Can Clark's picture of the human as "natural born cyborg" encompass the distribution of cognition into system-environment hybrids where the environment's agency is not trivial—that is, where it is more than a mere function of the autopoiesis of the bodybrain (no matter how flexible this latter might in fact be)? The answer would seem to be: to some extent at least. For in addition to positing an indirect, yet insuperable, coupling of organismic complexity and the complexity of contemporary technics, Clark's account of cognitive distribution directly implicates technics as a resource for the ongoing development of *both* the biologically housed cognitive system (the brain or bodybrain) and the larger cognitively distributed system (comprised of bodybrain and environmental scaffolding). Thus, even as the complexity of the environment must be domesticated from the standpoint of the organism's (impressively flexible) biological closure, it can—indeed must—be deployed as a resource, a source of alterity, for the correlated evolution of the brain and the technically distributed cognitive systems in which it ever-increasingly finds itself implicated today.

Returning now to our discussion of the irreducibility of environmental and technical complexity, we can see clearly how Clark's work authorizes (even if it doesn't itself develop) an important alternative to Hayles's call for a general leveling of boundaries and a turn away from neocybernetic conceptions of recursivity. Specifically, Clark's conceptualization of cognitive distribution as a model for human-machine mergers allows us to institute and maintain a differentiation between cognitive flexibility and biological closure that can recognize and capitalize on the increased technical complexification of the environment at the same time as it affirms the motif of closure, not once and for all, but multiply and differentially, at different levels of cognitive operation. The pains Clark takes to distance his understanding of openness from the scenario of machinic intrusion into the flesh thus begin to make sense as part of a larger argument for the displacement of the static binary of closure-openness in favor of dynamic functionalist differentiations. In this respect, Clark's account is notable for its effort *to preserve the specificity of* the divergent elements—"material brain, material bodies, and complex cultural and technological environments"—whose "looping interactions" generate "human thought and reason."[13] Rejecting any wholesale de-differentiation of boundaries, Clark's position instead urges a multiplication and complexification of boundaries—and specifically the boundaries constituting the living human being as a (multi-leveled) system—such that the biological flexibility of the human being can open up new cognitive dimensions, but only when correlated with the most creative, culturally and technologically

catalyzed, interactional possibilities. At the limit then, Clark's conceptualization authorizes a configuration of system-environment correlation such that
complexity in the latter, far from being first and foremost a problem that must
be mastered by system selectivity (as it is for Luhmann), can itself function as a
source for system operation or, more precisely, as a source for system change.
By positioning environmental complexity as a crucial source for the cognitive
operation and development of the human being understood both as embodied
brain and as element in a larger distributed cognitive system, Clark embraces
the introduction, at least in principle, of some measure of *heteropoiesis* into
system operation.

To theorize the positive contribution of heteropoiesis, we must, however,
move beyond Clark's emphasis on how biological flexibility manages to transform the complementarity of closure and openness; we must, that is, focus on
how cognitive function is actually shared among human *and* machinic components in actual instances of SEHS. For this task, let us turn to the work of French
"schizoanalyst" Félix Guattari, whose account of "machinic heterogenesis" taps
the promise of Varela's decoupling of closure from autopoiesis precisely in
order to open system function to the force of the heterogeneous technosphere.
Autopoiesis, Guattari suggests,

> deserves to be rethought in relation to entities that are evolutive and col-
> lective, and that sustain diverse kinds of relations of alterity, rather than
> being implacably closed in upon themselves. Thus institutions, like technical
> machines, which, in appearance, depend on allopoiesis, become ipso facto
> autopoietic when they are seen *in the framework of machinic orderings that
> they constitute along with human beings.* We can thus envision autopoiesis
> under the heading of an ontogenesis and phylogenesis specific to a mecha-
> nosphere that superimposes itself on the biosphere.[14]

His retention of the term "autopoiesis" notwithstanding, what Guattari is at-
tempting to theorize is precisely a deployment of closure beyond the narrow
boundaries of autopoietic closure—a deployment of closure in the service of
defining less stable, less unified, but nonetheless definitively bounded and consistent machinic orderings that largely coincide with what I have been calling
system-environment hybrids. Guattari makes clear the tradeoff involved in such
a softening of the motif of closure: the demand for unity must be tempered,
as must the stricture on (at least some) informational transfer across what
will henceforth become provisional, dynamically evolving boundaries: "Dia-
grammatic virtualities lead us away from Varela's characterization of machinic
autopoiesis as unitary individuation, without input or output, and prompt us

to emphasize a more collective machinism without delimited unity and whose autonomy meshes with diverse bases for alterity."[15] Part of my argument here is that such a picture—of a dynamic process involving multiple, provisional closures across the human-machinic divide—corresponds to the insight behind (though, to be sure, not to the concrete working out of) Varela's own decoupling of closure from autopoiesis.

For Guattari, as for Varela (and for me, too), this decoupling has an irreducible ethical dimension—it urges us to think and act in sync with the material complexification of the cosmos itself. Guattari privileges technical machines over all machinic orderings that impose a universal referent (capital, energy, information) precisely because they cut across material domains: technical machines, he insists, "are founded at the crossroads of the most complex and the most heterogeneous enunciative components."[16] When he says that "machines speak to other machines before speaking to man," Guattari clearly enunciates their role as mediators of environmental complexity. Machines are integral to any effort to impose some provisional closure, some fleeting reduction of complexity, on a world, a technosphere, increasingly characterized by relentless heterogenesis. As such, they are mediators for human co-evolution with the environment. And human cooperation with machines across system boundaries—cooperation or co-functioning with the alterity and nontrivial agency of technical complexification—is precisely how cognition gains purchase in the world today. Such cooperation or co-functioning, moreover, is absolutely crucial to the future of the human as a form of life inescapably coupled to its environment, for if the human is to retain its relevance, if not necessarily its centrality, in the face of the massive and massively accelerating complexification of the world, human beings must welcome the alterity of machines as a crucial source of connection to a world ever more difficult to grasp directly.

What Guattari simply calls "machines" (what I have termed system-environment hybrids) are operators of dynamically evolving, provisional closure that are at once necessary to decomplexify what would otherwise be an unassimilable chaos but that respect the alterity, indeed the "autonomy," of the world's material complexity. At the same time as they institute closure, and indeed, precisely to do so in their singular, alterity-respecting manner, machinic orderings draw on the extent of the world's complex intercrossings: every machinic ordering, Guattari claims, "through its various components, tears away its consistency by crossing ontological thresholds, thresholds of nonlinear irreversibility, ontogenetic and phylogenetic thresholds, thresholds of creative heterogenesis and autopoiesis."[17] Far from imposing the univocal figure of Being on the world, such a picture of human-machine correlation affirms not just the particular closures, the concrete machines that support it, but along with them, the alterity of material complexification itself: "The play of intensity within the ontological constellation is . . . a choice of being not only for itself but for all the alterity of the cosmos and for the infinity of time."[18]

The Autonomy of System-Environment Hybrids

Guattari's conceptualization of machinic heterogenesis expresses the ethical imperative we face today, the imperative to refuse the twin temptations posed by contemporary environmental complexity, on the one hand, simply to dissolve boundaries altogether (Hayles), and, on the other, to harden boundaries into a handful of durable autopoietic system types (Luhmann). However, Guattari doesn't himself grapple with the concrete difficulties involved in developing a viable model for the machinic orderings or system-environment hybrids that, arguably, comprise the predominant forms of organization (or dynamic, provisional closure) in our world today. To begin grappling with these difficulties, I propose that we return to the inspiration for Guattari's work and also for my own intervention here—namely, Varela's conception of autonomy, specifically in the mature, final form it takes once it has been decoupled from autopoiesis. What Varela provides is a commitment to continuity (not identity) across difference and, in the wake of this commitment, an ethical perspective capable of affirming the irreducibility of the human as a form of living even as it tracks the alterations inescapably imposed on this form of living by our world's ever-accelerating environmental complexification. To be blunt, what distinguishes a Varela-inspired account of today's SEHs is its commitment to the irreducibility of the human perspective in the face of its ever more complexly configured technical distribution.

Fleshing out such an account—an account that conjoins a commitment to human closure with the differential reiteration of closure across system levels—will, of course, require us to move beyond the resources of Varela's project. In what follows, I propose a genealogy for such an expansion that detours through the work of Castoriadis and Simondon. In work that intersects directly with Varela's research in biology, Castoriadis demonstrates the necessity to move beyond not just the closure-openness binary, but more specifically the level-specificity of the operation of autonomy: whereas autonomy of the living requires organizational closure, autonomy of psychic and collective beings requires openness to alterity, which is to say willingness to embrace precisely that which motivates system change. And with his conception of individuation as a necessarily incomplete process involving a concrete individual and the "preindividual" environment, Simondon links this openness to alterity—this embrace of heteropoiesis—directly to the technical complexification of the environment; on Simondon's account, not only does the environment necessarily possess nontrivial agency in *every* process of individuation, but to the extent that this environmental agency carries the "energy" that opens the unexpectable, it also plays the major role in all creative change.

Together, Varela's insistence on the integrity of the human and on continuity across divergent levels of being, Castoriadis's differentiation of levels of autonomy, and Simondon's privileging of the agency of the environment assemble the tools necessary for us to theorize, in a broadly neocybernetic mode, the operation of system-environment hybrids made necessary by the contemporary complexification of our technosphere. Rather than possessing institutional (that is, autopoietic) closure that cuts across the human (as Luhmann rather triumphally puts it), today's SEHs are created and dynamically evolve through what I would like to call *technical closure*, which is to say through far more provisional forms of closure that are resolutely functional in nature; that persist only so long as the distributed cognitive operations they characterize last; and that are minimal in the sense of having no significance other than their loose selection of relevant agencies without regard to the maintenance of more stable, narrowly defined, and pure system boundaries. As such, technical closure allows the environment to exert its agency independently of and beyond the demands of system autopoiesis (closure more narrowly defined as autopoietic closure), including the autopoiesis of the living human agents that form a robust, nondivisible or integral component of so many (though, to be sure, by no means all) SEHs in our world today.

Continuity: Varela

Varela stands apart from other neocyberneticians on account of his insistence on maintaining some form of continuity across system levels. From Varela's perspective, the human being is *at once* and *inseparably* a living being comprised of sublevels of autopoietic systems (the cellular and the immunological, following his five-part division) *and* a psychic and social being—that is, a living being whose activity of living happens through psychic and social events, in a psycho-social milieu. And while Varela recognizes that these levels can be isolated from one another in order to analyze them scientifically, his work forcefully underscores how any such isolation necessarily involves an abstraction from a consistency that characterizes the living (human) being as a totality and one that, crucially, encompasses its continuous, if ever-changing, relation with a complex environment.

As I see it, Varela's commitment to continuity across system levels has its origin in a certain methodological principle that dates to the very inception of the theory of autopoiesis—namely, the primacy accorded the operational perspective of the system over any observational vantage point. This primacy works to constrain the observations that can be made of a system: "A proper recognition of an autopoietic system as a unity requires that the observer perform an operation of distinction that defines the limits of the system *in the same domain in which it specifies them through its autopoiesis.* If this is not the case, he does not observe the autopoietic system as a unity."[19] This seemingly straightforward specification of proper observational technique—this injunction to delimit the system in strict accordance with its own operational closure—comprises what I would like to hold up as an ethical principle for neocybernetics: the principle of respect for system closure. Once it is broadened in conjunction with Varela's later decoupling of closure from autopoiesis, this ethico-ontological constraint might well be considered to be in itself a distinct form of autonomy (or at least a way of understanding such a distinct autonomy)—namely, the form of autonomy characteristic of the provisional (or technical) closures that constitute today's SEHS. Today's SEHS, that is, simply expand an ethical perspective originally developed for the purpose of respecting the consistency of the human as a form of the living (that is, as a being rooted in biological autopoiesis); in perfect analogy with human beings (and other living systems), SEHS will have to be observed from the standpoint of their operation, which is to say in a manner that respects the concrete technical closure(s) to which they owe their constitution.

This broadening of the ethical principle motivating Varela's work goes hand-in-hand with the discovery of continuity across systems levels, as Varela explains in the course of a 1995 interview with his friend, Cornelius Castoriadis:

> Life as an auto-constituting process already contains this distinction of a for-itself that, as Corneille [Cornelius Castoriadis] would say, is the source of that from which emerges the imaginary, capable precisely of giving meaning to what is only an array of physical objects. This rootedness of meaning in the origin of life is precisely where we find the novelty of this concept of autonomy, of autopoiesis. . . . What I have just said, that there is an excess of the imaginary that comes from this autoconstitution of the living, is one of the things I have learned in reading Corneille. And I would never have

dared speak of the imaginary at the origin of life if I had not found this type of continuity between biological phenomena at the origin of life and the social domain. Note that I say continuity and not identity.[20]

In Varela's description, it is the excess of the imaginary, coincident with the advent of living systems, that drives the spiral of self-organization ever upward. Indeed, it is the persistence of self-organization across divergent levels of organization that effectively calls for a rethinking of autonomy beyond autopoiesis. Doing justice to the fundamental continuity between the biological and the social domain will thus require a certain de-specification of autonomy, such that the latter comes to name the minimal conditions for a nonreductive recognition of system unity.[21] Generalized as a minimal principle of closure, autonomy forms the linchpin for an account of the human being as a multiply differentiated, complexly recursive hybrid of overlapping, though heterogeneous, systems and SEHS.[22]

While such an account remains beyond the resources of Varela's work, his effort to develop a five-level scheme of human emergence does take a small step in the right direction.[23] Specifically, it manifests a conviction that theorizing human agency in a complex socio-technical world like ours requires a recognition of continuity across levels and an embrace of multiple and heterogeneous forms of autonomy. More concretely, the turn to emergence allows for the preservation of consistency (what Varela calls the "unity" of the human) despite the heterogeneity of levels—and of types of autonomy—involved. Emergence, that is, adds to Varela's flexible and capacious conception of autonomy a global perspective *that is not external to the operation of the organism, that does not require an external observational vantage point* because an organism (and, more generally, a system) institutes its structural coupling with the environment in the very act of specifying its own unity, its interactional domain ceases being external to it, in the sense of requiring an observer's perspective; rather, the set of interactions an organism can maintain with the environment becomes, like the unity defining it, something that the organism itself specifies in the process of its self-organizing emergence. By integrating the interactional domain—what Varela calls the "surplus of significance" or, simply, the "world" (as against the "environment")[24]—into the multi-level, self-differentiated system of the organism, this dialectical conception of emergent selfhood *renders the interactional domain a factor in the ongoing evolution of the internally differentiated system of the organism.*

However, the agency thereby accorded the environment remains highly constrained. Indeed, if it doesn't serve simply to support the autopoiesis of the system in question (as it invariably does for Luhmann and for Varela's more delimited accounts of the immunological and cellular systems), environmental agency here functions exclusively to bolster cognitive capacities that are firmly rooted in the system's already established operationality. Indeed, the very rationale of Varela's differentiation of world ("surplus of significance") and environment is to separate out from the chaos of the latter precisely that part which is compatible with and supportive of the agency of the system.[25] True to his claim that the environment "provides an 'excuse' for the neural 'music'" of the organism,[26] Varela remains unable to deploy the agency of the environment nontrivially, which is to say for itself.

Discontinuity: Castoriadis

In the course of his differentiation of "world" from "environment," Varela specifies that the "surplus of signification" it yields is due to the "perspective provided by the global action of the organism." The surplus of significance, that is, results from the operation of the organism as a "for-itself" and thus properly concerns what Varela glosses, with due recognition of its potential for semantic abuse, as the organism's "*imaginary* dimension."[27] Undoubtedly a reference to the work of Castoriadis, this invocation of the imaginary opens a rift in Varela's biological holism, at least if we follow Castoriadis in differentiating (though not divorcing) the imaginary from the biological stratum. Specifically, it calls on us to expand the foundation of emergence beyond the biological domain assumed by Varela.

Such an expansion has the salutary effect of according a certain modicum of (nontrivial) agency to the environment: whereas for Varela (at least for the Varela of *Autopoiesis and Cognition*) it is the biological (or autopoietic) organism that determines how (what part of) the environment can matter, for Castoriadis, by contrast, the environment has the power to impel changes in the system (organism); accordingly, whereas Varela's organism recognizes only that part of the environment that supports system (organism) autopoiesis (hence the distinction between "world" and "environment" as such), on Castoriadis's account, the environment matters, above all, *as a source of alterity* for a system. This is a crucial difference, for it starkly divorces Castoriadis's conception of higher-order (social) autonomy from any attempt to perpetuate autopoietic autonomy at higher levels, of which the most notable is certainly that of Luhmann. Thus, in order for societies to be autonomous—which Castoriadis defines as "determining their own laws"—they must be capable of directing their own transformation, a process that requires an openness to, indeed an embrace of, (environmental) alterity. Bluntly put, higher-order (societal) autonomy requires a break with autopoietic autonomy. Emboldened by Varela's remark to me that Luhmann was the worst thing to have happened to him, I want to suggest that a full-scale development of the Varelean systems perspective at the societal level would follow the Castoriadis (and not the more well-trodden Luhmann) fork toward the reconciliation of autonomy and alterity.[28]

Because of its nonreducibility to the organic, the imaginary entails a process of world constitution as something other than a trivial support for the organism's (system's) autopoiesis. Castoriadis expresses this clearly when he claims that "radical imagination . . . is what makes it possible for any being-for-itself (including humans) to *create for* itself an own [or proper world] (*eine Eigenwelt*) 'within' which it also posits itself. The ultimately indescribable X 'out there' becomes something definite and specific *for* a particular being, through the functioning of its sensory and logical imagination, which 'filters,' 'forms,' and 'organizes' the external 'shocks.'"[29] While this description seems entirely of a piece with Varela's superficially similar description of an organism's *informing* of brute materiality, the absolutely crucial difference is that for Castoriadis, what the system reaches out to and in-forms is *radically other to it*. While this alterity becomes the basis for the very creation of the system (and thus necessarily ceases to be *radically* other), far from being a source for ongoing reproduction as it is on Varela's account, it is the source for creation, for the creation of myriad possible closures that, no matter how provisional, each mark the advent of something radically new.[30]

In Castoriadis's work, this extra-biological dimension of the process of informing the environment is the source for an explicit break, at higher levels of organization, with the autopoietic form of closure and, accordingly, for a decoupling of autonomy (not so much, as Castoriadis would have it, from closure per se as) from autopoiesis.[31] Rather than guaranteeing the self-perpetuation of the living organism as the particular living organization that it is and must continue to be, autonomy involves an opening up to alterity and a correlative opportunity to alter the organization of the self: "Autonomy," Castoriadis insists, "is not closure but, rather, opening: ontological opening, the possibility of going beyond the informational, cognitive, and organizational closure characteristic of self-constituting, but *heteronomous* beings. It is ontological opening, since to go beyond this closure signifies altering the already existing cognitive and organizational 'system,' *therefore* constituting one's world and one's self according to *other* laws, *therefore* creating a new ontological *eidos*, another self in another world."[32] His thoroughgoing antipathy to the term "closure" notwithstanding, it seems to me that Castoriadis is here simply extending the argument, initiated by Varela in his own decoupling of autonomy from autopoiesis, that closure takes different, non-autopoietic forms at higher levels of organization.

To theorize the extra-biological dimension of the imaginary, Castoriadis introduces the concept of leaning-on. An expansion of Freudian *anaclisis*, this notion describes the capacity of the *psyche* to break from the natural stratum while nonetheless continuing to find anchoring in it; in Castoriadis's work, this operation of splitting through leaning-on will be repeated at subsequent levels of organization, up to and including the societal. Indeed, the concept of leaning-on allows Castoriadis to detach the imaginary from the biological and to correlate it with the role of institution, such that all world constitution (the imaginary in-forming of an X), from the most primitive to the most complex, necessarily involves an intrinsic institutional element. Rather than being biologically determined, as it is for Varela, emergence is thus properly imaginary and, as such, is *constrained* but not determined by the biological stratum: "To say that the institution of society *leans on* the organization of the first natural stratum [defined as the 'set and stable organization of a part of the world homologous to the organization of humans as simple living beings'] means that it does not reproduce or reflect this organization, is not *determined* by it in any way. Instead, society finds in it a series of conditions, supports and stimuli, stops and obstacles."[33] Emphasizing that creation *ex nihilo* always implies constraints (biological above all, but also "historical" and "intrinsic"),[34] Castoriadis sharply differentiates his conception of emergence through radical imagination from Varela's conceptualization of emergence *ex bios* and from all models that postulate a continuity of emergence, including "systems theory," "self-organization," and "order from noise."[35] Like all continuist accounts—even one that, like Varela's, exploits the nonseparability of the local and the global perspectives—these theories can at best furnish the necessary—but never the sufficient—conditions for emergence.[36]

With its emphasis on *ex nihilo* creation through opening to the alterity of the environment, Castoriadis's work allows us to modulate between continuity and discontinuity, to think the latter without simply abandoning the former. Crucially, it is this capacity to implicate discontinuity within continuity that underwrites the theorization of the human as an accumulation of multiple and heterogeneous closures. Because it furnishes the force behind all acts of creation (which, contrary to Castoriadis's desire, I would associate with acts of non-autopoietic closure), the imaginary operates the re-individuation of the natural stratum (the biological organism) such that this latter becomes encompassed within a broader, internally complex, multiple, and heterogeneous process of individuation, what we might well call (combining the terminology of Castoriadis and Simondon) the institutional individuation of the human being.[37]

Discontinuous Continuity: Simondon

It falls to Simondon to furnish the theoretical underpinnings for such an account of the complex individuation of the human. In work richly deserving of recognition in English-speaking contexts, Simondon develops a complex theory of individuation that spans the continuum from the physical to the collective domain and that takes as its first principle the essential incompleteness of all processes of individuation.[38] At all levels from a crystal to a human society, individuation always occurs in conjunction with a "preindividual" domain, a domain of excess or alterity, which Simondon defines as "a certain inheritance" of the individual "animated by *all* the potentials that characterize it."[39] The preindividual domain furnishes a source of excess or alterity that doesn't simply perpetuate processes of individuation but that—far more crucially—ensures their almost continuous self-modification, their ongoing emergence, or continuous creativity.

By conjoining individuation and the preindividual, Simondon articulates a concept of ontogenesis or emergence that can be viewed as expanding insights from Varela and Castoriadis toward the end of theorizing the robust agency of the environment, not as a support for the system (individual) but for itself. As we shall see, such theorization brings together a transductive account of continuity and discontinuity, a fluid shifting of scales between individual and collective agency, and an attention to the crucial role played by technics in higher-order processes of individuation.

We can get a clear sense of what Simondon brings to the neocybernetic table simply by contrasting his concept of the preindividual with Varela's and Castoriadis's broadly similar concept of the environment. As we noted above, for the latter theorists, the environment must be "in-formed" or reduced to a "surplus of significance" in order to be meaningful to an organism or system. Such a requirement emphatically has no role in Simondon, where the preindividual operates independently of any preliminary accommodation to an individual and thus as an inescapable source of alterity for the individual. This difference has important consequences for our understanding of emergence, for whereas Varela connects emergence to the co-functioning of local and global perspectives *of a system*, Simondon links emergence—which he defines as the passage to a higher level of individuation, one that encompasses the system and

the (preindividual) environment—to a more encompassing recontextualiza-
tion of individuation or, more exactly and more simply, to a (new) reindividuation of the (same) preindividual charge of reality. For Simondon, in short,
it is not simply the global perspective *of the organism*—a perspective tied to
the organism's specification of a world—that informs the bootstrapping of
identity from level to level; rather, the upward spiral of individuation is driven
by two important conditions: the nonidentification of individuation with any
form of individual (physical, biological, psychic, or collective) and the coupling
of individuation with the entire environment as a source of "preindividual,"
"metastable" potential. Together, these conditions ensure that emergence *qua*
individuation involves a recursivity that is not driven solely or primarily by the
organism's demands but that instead draws from the global *situation*—the preindividual as potential—within which all individuations necessarily occur.

It is crucial to grasp the distinction being made here: global situation (Simondon) differs from global perspective (Varela) *because it is not relative to the
organism alone (or, more exactly, to the organism plus its world), in isolation from
the environment.* Put otherwise, if the global situation is a global perspective,
it is not a perspective *of the organism* but a perspective *on the entire process of
individuation* of which the organism is only one part—a perspective, in short,
that situates the organism within the broader context of the preindividual.
Here, in other words, we have a perspective that coincides with what I have
been calling the system-environment hybrid, a mixed form of agency where an
individual (in most cases a human being) has cognitive agency in conjunction
with an environment that it does not control and cannot reduce.

Indeed, with his insistence on the primacy of relation and the resulting
double relationality of individuation, Simondon adds a crucial component to
the concept of the SEH—namely, continuity across the potentiality-actuality
divide.[40] Because it implicates both individual and preindividual, individuation
is relational in two distinct senses and within two different frames of reference.
As the relation of the individual and its "associated milieu" (more or less synonymous with Varela's "world"), individuation occurs within a single level or
"order of magnitude" and is, as such, entirely actual; such a relation is roughly
equivalent to what Varela calls "structural coupling." As the relation of the individual and the preindividual, by contrast, individuation occurs across distinct
orders of magnitude and thus concerns the potential prior to or "beneath" its
actualization. Not only does Simondon's conceptualization of the equiprimacy
of individuation and relation thus situate "structural coupling" within a more
encompassing ontogenesis, but it also accords the preindividual—that is, the
environment not already coupled to the individual system—a mode of *potential*

efficacy rooted in what he calls "metastability."[41] On Simondon's account, in short, agency encompasses the environment as a whole and as a source of potential that may or may not be actualized and, indeed, *that need not be actualized by the organism* in order to impact its individuation. As against Varela's thinking of continuity from out of the biological, in which continuity is and can only be between actualized individuals, Simondon's account of individuation ascribes the continuity underlying emergence to (preindividual) potential.

For Simondon no less than Varela and Castoriadis, the human is a complex entity comprised of multiple systems and agencies, one that is developed on the basis of emergences or leanings-on that cut across levels of organization and defy the purity of neocybernetic theorization narrowly conceived. It is, however, only with the introduction of Simondon's complex, doubly relational account of individuation that we can grasp the consistency of the human in all of its complexity. In large part, this is due to the central role Simondon grants technics: in his work, which includes a study of the ontogenesis of technical objects, Simondon situates technics between the preindividual and actualized individuation.[42] As such, technics expands the scope of the environment's agency (and Simondon's focus on technics, we should note, has much to do with his appreciation for environmental agency). Specifically, technologies mediate the preindividual in a way that facilitates collective individuations that are not simply agglomerations of individuals but new, properly collective individuations of preindividual potential. Simondon refers to this kind of collective individuation as *transindividuation*, and it is significant that he associates it consistently with the "essential" correlation of human being with technics. On Simondon's account, transindividuation comprises the ultimate stage of the individuation of the human; to realize their full potential for individuation, human beings must not only deindividuate themselves enough to reindividuate as the transindividual, but they must also rely on technics to do so.

By correlating transindividuation and technics, Simondon furnishes a paradigm for encompassing the environment—indeed, the highly advanced technosphere of today—as a nontrivial element of cognitive action. As "symbols" that "express" the preindividual reality attached to the subject, technical objects transform this preindividual potential into an actualized excess that no longer corresponds to the individual–associated milieu coupling but that facilitates their individuation as a collective being. Technical objects transform the preindividual reality associated with the living individual into an actualized source of energy that, as Simondon puts it, "*surpasses the individual while still prolonging it.*" "The transindividual," Simondon goes on to conclude, "is not exterior to the individual and yet it detaches itself to some extent from the individual."[43]

Not only do technical objects make possible a surpassing of the individual, but they also transform the preindividual reality that exceeds the individual into the basis for a new individuation, a new actualizing of this excess. Technical objects thus actualize what Bernard Stiegler has called *epiphylogenesis*, that dimension of human evolution that occurs through the exteriorization of the living in tools, language, archives, and institutions.[44] As such, technical objects convert the preindividual excess associated with the living individual into a transindividual excess associated with that crucial dimension of the human that is radically exterior to its biological or zoological individuation.

The mediation performed by technical objects on Simondon's account—the actualization of the preindividual potential as a source for collective reindividuation—furnishes the means for the living (human) organism to experience *heteropoiesis*, to be impacted by the environment without the latter first undergoing a selective reduction keyed to the organism's constitutive organization. Technical mediation thus furnishes an account of the coupling of the human and the technical that can specify precisely how the technical scaffolding of human cognition (in the largest and most embodied sense of the term) holds some agency in the development of the human being, for what we encounter in instances where the environment is coupled to the human as a source of potential that is simultaneously *necessary* and *heterogeneous* is precisely the potential for the human to change in ways that go beyond, if not indeed against, its constitutive closure and the scope of interactions associated with it. Without abandoning the continuity across the divergent levels of individuation—the continuity *of the living*—Simondon's account allows us to think the human capacity for radical creation and to grasp the specific and crucial role played by technics not simply as part of its cause or motivation, but also as its irreducible support, its medium.

Ultimately, then, the value of Simondon's broadly neocybernetic conception of individuation for thinking the contemporary phase of our human technogenesis stems from its sublation of a differentiation—of systemic from environmental complexity—that is crucial to Luhmannian and von Foersterian neocybernetics. In concert with contemporary cultural critics like Hayles, Guattari, and Clark, Simondon recognizes and values the irreducible and increasing complexification of the environment and forcefully embeds all forms of cognitive agency within it. And yet, in contrast to these critics, Simondon associates the complexification of the environment with a parallel complexification of individuation understood as a process that maintains some form of closure, some constitutive organization, even as it remains associated with, indeed informed by, the environment *qua* heteropoietic potentiality. For Simondon, in short, environmental complexification drives individuation from the get-go. Coupled with the reality that, today, environmental complexification *is* technical complexification, this fundamental commitment renders Simondon's work perfectly suited for thinking the contemporary technogenesis of the human and for expressing the ethical irreducibility of a minimal neocybernetic commitment that must form the core of any such thinking.

Notes

1. Following the felicitous concept introduced by anthropologist Edwin Hutchins in his study of navigation systems and practices upon large seafaring ships, *Cognition in the Wild*. The sources for the epigraphs are as follows: von Foerster, "Molecular Ethology," 152; corrected from the version cited in Varela, "Introduction: The Ages of Heinz von Foerster," xv; Guattari, "Machinic Heterogenesis," 17; Latour, "Is Re-Modernization Occurring?" 38; emphasis in original.
2. Luhmann, *Social Systems*, 29.
3. Ibid., 27; emphases in original.
4. Thrift, "Remembering the Technological Unconscious by Foregrounding Knowledges of Position."
5. The terminology here is, of course, Luhmann's; though largely beside the point here, I side with Varela in claiming that organismic and suborganismic closure *is* cognitive and yields meaning for the system in question.
6. Varela, "Introduction: The Ages of Heinz von Foerster," xv.
7. See von Foerster, "Molecular Ethology," 152.
8. Hayles, *My Mother Was a Computer*, 280.
9. Clark, *Natural Born Cyborgs*, 7. "What really matters," Clark continues, "might be just the *fluidity* of the human-machine integration and the resulting *transformation* of our capacities, projects, and lifestyles. It is then an empirical question whether the greatest usable bandwidth and potential lies with full implant technologies or with well-designed nonpenetrative modes of personal augmentation. With regard to the critical features just mentioned, I believe that the most potent near-future technologies will be those that offer integration and transformation *without* implants or surgery: human-machine mergers that simply bypass, rather than penetrate, the old biological borders of skin and skull" (24); emphases in original.
10. Ibid., 5.
11. Ibid., 10; emphases added.
12. Ibid., 85.
13. Ibid., 11.
14. Guattari, "Machinic Heterogenesis," 17; emphasis added.
15. Ibid., 18.
16. Ibid., 22.

17. Ibid., 24.

18. Ibid., 26.

19. Maturana and Varela, "Autopoiesis," 109; emphasis added. "An explanation is always a reformulation of a phenomenon showing how its components generate it through their interactions and relations. Furthermore, an explanation is always given by us as observers, and *it is central to distinguish in it what pertains to the system as constitutive of its phenomenology from what pertains to its domain of description,* and hence to our interactions with it, its components and the context in which it is observed. Since our descriptive domain arises because we simultaneously behold the unity and its interactions in the domain of observation, notions arising in the domain of description *do not pertain to* the constitutive organization of the unity (phenomenon) to be explained" (ibid., 75; emphases added).

20. Castoriadis, "Entretien Cornelius Castoriadis et Francisco Varela," 102–3; my translation.

21. Such a despecification is central to Varela's first major solo work, *Principles of Biological Autonomy*: "What other autonomous systems have in common with living systems is that in them, too, the proper recognition of the unity is intimately tied to, and occurs in the *same* space specified, by the unity's organization and operation. This is precisely what autonomy connotes: assertion of the system's identity through its functioning in such a way that observation proceeds through the coupling between the observer and the unit in the domain in which the unity's operation occurs" (Varela, *Principles of Biological Autonomy*, 54; emphasis in original).

22. This decoupling comprises Varela's response to attempts to generalize the application of autopoiesis to nonliving human systems, such as social institutions. Calling these "category mistakes" because they "confuse autopoiesis with autonomy," Varela specifies autopoiesis, which is henceforth defined by its restriction to relations of production of some kind and its reference to topological boundaries and to conceptualize autonomy, such that it becomes available to characterize a rich variety of systems: "The relations that characterize autopoiesis are relations of *productions* of components. Further, this idea of component production has, as its fundamental referent, chemical production. Given this notion of production of components, it follows that the cases of autopoiesis we can actually exhibit, such as living systems or model cases like the ones described in Chapter 3, have as a criterion of distinction a topological boundary, and the processes that define them occur in a physical-like space, actual or simulated in a computer" (Varela, *Principles of Biological Autonomy*, 54; emphasis in original).

23. In "Organism" (1991), Varela differentiates five levels of selfhood that are correlated through "reciprocal causality" between the local and the global. Varela enumerates five such levels (or forms of identity) together with five correlative "regional selves": (1) cellular identity (biological self); (2) immunological identity (bodily self); (3) behavioral identity (cognitive self); (4) personal identity (sociolinguistic self); and (5) social identity (collective self). On this model, the identity of an organism is highly differentiated in the sense that it encompasses multiple levels of development,

each of which enacts a process of self-constitution. Though each of these "regional selves" comprises an emergent level of development, from the most basic cellular identity common to all living organisms to the most specialized personal and collective identity characteristic of human organisms, they each display the same pattern of continuous self-specification of systemic boundaries in the face of constant perturbations from the environment. Varela's willingness in this essay to embrace "recent notions of emergent properties in highly distributed, modular systems" follows directly upon his insight, formulated as early as 1979 in *Principles*, into the necessarily distributed scope of human cognition: "The act of understanding is basically *beyond our will*, precisely because the autonomy of the social and biological system we are in goes *beyond* our skull, because our evolution makes us part of a social aggregate and a natural aggregate, which have an autonomy compatible with, but *not reducible* to, our autonomy as biological individuals" (Varela, *Principles of Biological Autonomy*, 276; emphases in original).

24. Varela's distinction between world and environment and the correlated notion of a "surplus of significance" permits a folding of the interactional domain into the system. Whereas "environment" designates the environment *of* the living system as viewed from the observer's perspective, "world" denotes the environment *for* the system, which, he specifies, "is defined in the same movement that gave rise to its identity and that only exists in that mutual definition" (Varela, "Organism," 85). Because it determines a system's interactional domain, world comprises an "exteriorization" or definition of "what remains exterior to [the system as a unity]" that can "only be understood . . . from the 'inside.' " "The autopoietic unity," Varela argues, "*creates a perspective* from which the exterior is one, which cannot be confused with the physical surroundings as they appear to us as observers" (85; emphasis in original). Varela describes the difference between environment and world as a "surplus of significance" produced by the organism's perspective and stresses not only that it "haunts the understanding of the living and of cognition," but also, and crucially, that it lies "at the root of how a self becomes one" (86). It is the surplus of signification that infolds a world into the system's perspective; since it comprises the product of the dialectical functioning of the organism, the surplus of signification forms a part of its self-specification. "What is meaningful for an organism," Varela emphasizes, "is precisely given by its constitution as a distributed process, with an indissociable link between local processes where an interaction occurs (i.e., physico-chemical forces acting on the cell), and the coordinated entity which is the autopoietic unity, giving rise to the handling of its environment without the need to resort to a central agent" (86). At the biological level, identity is maintained through the organism's continuous interaction with the environment, which functions *both* to specify the organism's boundaries *and* to carve out its world.

25. In a development more specific to the theorization of emergence, the opening of system operation to modification instigated by its "world" compels Varela to complicate the organism's constitutive dialectic, to double the "dialectics of the self" with a "dialectics of knowledge." This doubling positions the integration of the

interactional domain as the correlate of the coupling or "bootstrapping" of local and global perspectives that comprises identity at the various levels of system operation. "The key point," Varela specifies, "is that the organism brings forth and specifies its own domain of problems and actions to be 'solved'; this cognitive domain does not exist 'out there' in an environment that acts as a landing pad for an organism that somehow drops or is parachuted into the world. Instead, living beings and their worlds of meaning stand in relation to each other through *mutual specification* or *co-determination*" (Varela, "Organism," 103; emphasis in original).

26. Varela, *Principles of Biological Autonomy*, 276.

27. Varela, "Organism," 88.

28. Varela made this rather casual remark to me during a break at the conference honoring Castoriadis held at Columbia University in 2000.

29. Castoriadis, "Radical Imagination and the Society Instituting Imaginary," 326; emphases in original.

30. Castoriadis extends this structure all the way down, as it were, such that even the most primitive forms of organic self-organization—for example, the cellular—take place through such an assumption of radical alterity by the activity of the radical imagination: "What does a proper world signify? There is necessarily each time—at least as soon as one reaches the cellular level—presentation, representation, and a bringing into relation [*mise en relation*] of that which is represented. Certainly 'there is' something 'outside,' there is X. But X is not information, as its very designation here indicates. It 'informs' one only of the following thing: that 'there is.' It is mere shock, *Anstoss* (we shall return to this). As soon as anything more could be said about it, it would have already entered into the play of 'subjective' determinations; and ultimately, even this emptied, eviscerated limit case of determination that we are calling 'there is' is not exempt from the following question: *For whom* is there something? Nature contains no 'information' waiting to be gathered. This X becomes something only by *being formed* (in-formed) by the for-itself that forms it: the cell, immune system, dog, human being, etc. in question. Information is created by a 'subject'—obviously in *its own* manner of doing so" (Castoriadis, "The State of the Subject Today," 145; emphases in original).

31. Whether and to what extent this conception of the imaginary also requires a break with autopoiesis at lower levels of organization is beyond the scope of our concern here. See the citation in the note just preceding.

32. Castoriadis, "The Logic of Magmas and the Question of Autonomy," 310; emphases in original. As the radical capacity to give oneself one's own laws, autonomy results from the emancipation of the radical imaginary from biological and social-historical closure: "The autonomy of the individual consists in the instauration of an *other* relationship between the reflective instance and the other psychical instances as well as between the present and the history which made the individual such as it is. This relationship makes it possible for the individual to escape the enslavement of repetition, to look back upon itself, to reflect on the reason for its thoughts and the motives of its acts, guided by the elucidation of its desire and aiming at the truth.

This autonomy can effectively alter the behavior of the individual, as we positively know. This means that the individual is no longer a pure and passive product of its psyche and history and of the institution. In other words, the formation of a reflective and deliberative instance, that is, of true *subjectivity*, frees the radical imagination of the singular human being as source of creation and alteration and allows this being to attain an effective freedom" (Castoriadis, "Power, Politics, Autonomy," 165; emphases in original).

33. Castoriadis, *The Imaginary Institution of Society*, 234; emphases in original.

34. Castoriadis, "Radical Imagination and the Society Instituting Imaginary," 333–35.

35. Castoriadis, "Power, Politics, Autonomy," 145.

36. "We live in a world of colors that we create, but that we do not create entirely arbitrarily because they correspond to something, namely to the shocks that we receive from the external world. And this creation cannot be reduced to the simple reassembling of a host of local things. Indeed, the fact that a grouping of objects and their local connections functions as conditions leads to this idea, in my opinion totally elementary but astonishingly forgotten in this discussion, of the distinction between necessary and sufficient conditions. In order for the Greeks to create the *polis*, democracy, philosophy, scientific method, etc., there were a host, indeed an infinity, of necessary conditions . . . for example, Greek mythology, which is a necessary but not sufficient condition; there is an affinity of meaning but something else is required to create the *polis* and the other creations. To be yet more precise, creation never takes place *in nihilo* nor *cum nihilo*; as a form, it is *ex nihilo*. That is the *hic*, the bottom line, and it is for that reason that I believe non-linear mathematics can at best only furnish an *ex post* description of [a radical creation]" (Castoriadis, "Entretien Cornelius Castoriadis et Francisco Varela," 114–15; emphasis in original).

37. "Within this world an important place must always be found for the first natural stratum, whose being and being-thus (for humans as living beings) is the condition for the existence of society. But at the same time, this stratum is never, and could never be, taken up simply as such. What belongs to it is taken up in and through the magma of significations instituted by society, and in this way it is transsubstantiated or ontologically altered. It is altered in its mode of being inasmuch as it exists and exists only by reason of its investment by signification. It is also altered in its mode of organization and cannot help but be altered in this way. For not only is the mode of organization of the world of significations not the ensemblist mode of organization of the first natural stratum, but also, from the moment that everything has to signify something, this ensemblist organization does not, as such, answer the question of signification, and ceases to be an organization, even an ensemblist one" (Castoriadis, *The Imaginary Institution of Society*, 235).

38. For an introductory essay in English, see Gilbert Simondon, "The Genesis of the Individual."

39. Simondon, "The Genesis of the Individual," 306; emphasis added.

40. Simondon's insistence on the primacy of relation (or the equiprimacy of relation and individuation) forms a crucial aspect of his characterization of the preindividual as

real: "Individuation and relation are inseparable; the capacity for relation is a part of being, and enters in its definition and in the determination of its limits: there is no limit between the individual and its activity of relation. Relation is the contemporary of being; it forms a part of being both energetically and spatially. Relation exists simultaneously with being in the form of the field, and the potential that it delimits [*définit*] is real, not formal. Being in a potential form does not mean that energy does not exist" (Simondon, *L'Individu et sa genèse physico-biologique*, 141).

41. By characterizing the potentiality of the environment as "metastability," Simondon differentiates it from the virtual on account of its peculiar actuality; urging us to refrain from conceiving "preindividual reality" as "pure virtuality," Simondon insists that we see it as "veritable reality charged with potentials that are actually existing as potentials, that is to say, as the energy of a metastable system" (Simondon, *L'Individuation psychique et collective*, 210).

42. Simondon, *Du mode d'existence des objets techniques*.

43. Simondon, *L'Individuation psychique et collective*, 156.

44. Stiegler, *Technics and Time*. Vol. 1: *The Fault of Epimetheus*.

Self-Organization and Autopoiesis

NIKLAS LUHMANN

TRANSLATED BY HANS-GEORG MOELLER WITH BRUCE CLARKE

Editors' note: We give here a translation from Luhmann's *Einführung in die Systemtheorie*, the transcript of a lecture series Luhmann gave at the University of Bielefeld during the winter term 1991–92. The book was published four years after Luhmann's death, on the initiative of its editor, who slightly edited the recorded materials and added the footnotes. The text reproduces the oral characteristics of a lecture addressed to newcomers to systems theory within an academic context. In the following discussion, as an example of the self-organization of structures within autopoietic systems, Luhmann circles around the phenomenon of language acquisition. Language cannot be inserted into persons as an "input" by which they can then produce language as "output." Rather, in the co-production of psychic and social systems, linguistic structures and operations *self*-organize (if at all) due to psychic-systemic constructions of communicative events in their environment.

Self-organization and autopoiesis are two distinct concepts that I deliberately keep apart. Nevertheless, both are based on the theorem of operational closure—that is, both are founded not only on a difference-theoretical but also on a principally operational concept of the system. This is to say that the system has only its own operations at its disposal. In the system there is nothing else than its own operations, and these operations serve two different purposes.

The first purpose is the formation of its own structures. The structures of an operationally closed system are necessarily established by its own operations. Put differently, there is no importation of structures from elsewhere. This is called self-organization. Second, the system has only its own operations at its disposal to determine the historical state, or, if you will, the temporal presence, from which everything else has to emerge. With respect to the system, presence is determined by its own operations. For instance, what I have said just now is the point from which I have to proceed when I consider what I can say further. What I am thinking just now, what happens in this moment in my consciousness, what I perceive, is that which constitutes the starting point for the conceivability of further perceptions. I know that I am here in this room, at this place, and if I were to make erratic jumps, I would have to consider the possibility that I took some drugs and am therefore unable to achieve the normal continuity of perception that would help to explain the occurrence of surprising events. We are dealing, then, with two facts: first, with "self-organization" in the sense of a creation of a structure by means of a system's own operations, and, second, with "autopoiesis" in the sense of the determination of a state that makes further operations possible by means of the operations of the same system.

For a start, I would like to say something about self-organization. It is perhaps best to begin with the realization that structures within an operational theory are effective only at the moment at which the system operates. Here you can see a distance from classical conceptions because this contradicts the conception that structures are that which is persistent while processes or operations are that which fades away. In this theory, the structures are relevant only in the present. They can be made use of, if one may say so, only when the system operates, and anything that has happened at some other time or will happen at some other time is either in the past or in the future but not current. All descriptions of structures that are extended in time—and thus everything that one sees if, for instance, one sees that we are having this lecture always at the same time and always in the same classroom with perhaps not always the same people present but in any case with always the same lecturer—are related to structures that in turn demand an observer for whom the same is true—that is, who observes this only when he observes it, that is to say, when he is active or in operation. No matter if one thinks of the system in operation or relates the operation to the observation of other operations, everything is relative to a theme of simultaneity, presence, or the current moment. The system has to be in operation in order to be able to use structures.

Thus, if one now wants to know what a structure is, what a psychological structure is, how one would characterize persons, how one would describe them, how one would depict their habits, or, if one thinks of social systems, how one describes a university—one also sees that the description is in turn the description of an observer. One identifies the structures, but one does it only when a system that does it does it. This implies that the descriptive marking of structures is completely relative to a system's operations. And accordingly, this also means that projections into the past, and thus the falling back on the past, and, similarly, anticipations of the future, have to be adjusted to this theory. The structure is always only effective at the current moment, and the past data that are used in the moment are related to the actualization of projections into the future. In the present, and only in the present, there is the coupling of that which is commonly called memory with that which is normally called an expectation or a projection—or, if one thinks of actions, a goal.

Memory is not a stored past. The past is past and can never become current again. Memory is more a sort of consistency test, for which it is typically unnecessary to remember when one has either learned or not learned something particular. When I speak German right now, I do not have to know when I learned this language and how it came to this or when I used particular words, such as "autopoiesis," for the first time or when I read them for the first time. With respect to that which one aims at for the future in the context of expectations, anticipations, goals, etc., decisive is that which is currently on call, the current test of the breadth of the availability of structures, if you will. So far, this is a thoroughly pragmatist approach. There is a connection between the theory of memory, on the one hand, and a pragmatic orientation toward the future, on the other hand, which is always nicely narrowed down so that one could perhaps also say that memory is nothing but a perpetual consistency test of different bits of information with respect to specific expectations, be it that one intends something, be it that one is afraid of something, be it that one sees something coming and wants to react to it. This theory of memory does not suppose any kind of storehouse, which neurophysiology seems to confirm.[1] In the nervous system one does not find a past that would be stored in particular nerve cells, but one finds a cross-checking, a testing of various routine habits at particular occasions at particular moments.

Therefore the concept of expectation suggests itself to be used as the basis for the definition of structures. Structures are expectations in relation to the connectivity of operations, be it of mere experience, be it of action; and expectation in a sense that does not have to be understood as subjective, despite a criticism of this concept of expectation that accuses it of subjectivizing the notion of structure. However, the tradition of the concept of expectation is much older and not necessarily geared toward psychic structures. In the 1930s the concept of expectation was introduced to increase the complexity of rigid input-output relations or stimulus-response models—for instance, in order to be able to imagine that stimulus and response are not within fixed relations toward one another but are rather controlled by the expectations of the system. One can identify a stimulus only if one has specific expectations. One searches the terrain within which one perceives, within which one receives stimuli, with respect to expectations that one normally has or habitually supposes in a particular situation. This is also where the notion of "generalized" expectations (G. H. Mead) comes from. Initially this meant a break with the older behaviorist psychology, but then it was taken over by social theory in the sense that roles are bundles of expectations and that communication expresses expectations independently of what persons respectively think by themselves. Expectations thus express the future aspect of meaning that can either be given psychically or within social communication.

Because this theory defines the concept of structure on the basis of expectations, the subject-object distinction is unimportant for it. In Bielefeld, we once had a discussion with Johannes Berger, who criticized expectation as a subjective concept, and thus as one impractical for sociologists, who are more interested in objective social structures.[2] If you have taken courses in structural analysis, you will also probably have the impression that structures are objective facts that can be established statistically, or Marxistically, or however, without having to determine in a relative manner what individual persons respectively think. But in my approach to systems theory, you will see that I try to leave this subject-object distinction behind and replace it with the distinction between, on the one hand, the operation that a system actually performs when it performs it, and, on the other hand, the observation of this operation, be it by this system or be it by another system.

The concept of expectation then no longer contains any subjective component. Instead, the concept of expectation asks the question: How do structures achieve the reduction of complexity without reducing the system to merely one capacity? How can one imagine that a system disposes of a rich variety of structures—of language, for instance—without being limited by the choice of this or that sentence, and thus without immediately losing again this variety of structures? To the contrary, often structures are established without the intention or ability to determine the situation in which one will make use of them. The concept of structure has to explain why the system does not shrink when it continuously has to make decisions, when it continuously has to perform this or that connective operation—why it does not shrink but perhaps grows and gains complexity although it is continuously forced to reduce complexity. The more possibilities a system has—think again of language as the extraordinary example that demonstrates this—the more selective each single sentence and the less stereotypical speech is. This can be seen, for instance, when comparing social strata, if one compares the somewhat stereotypical form of speech of the lower strata—particularly in England—with the more elaborate form of language in the higher strata. An elaborate language finds the right word in every situation and is often capable of expressing itself more appropriately, especially when the structure is very elaborate and highly complicated, than when there is only a restricted vocabulary at hand. This has to be grasped with the concept of structure. Herein lies the achievement of joining high structural complexity with operational capability. Then you will also see why the shift from trivial to nontrivial machines is important. Trivial machines dispose only of those operations that are determined by the program and the input, whereas structurally complex systems can themselves effect the learning of how to match situations and, at the same time, they can, if you will, take their own thoughts or their own communicational habits into account so that these systems have a far richer repertoire of possibilities for action. The problem consists in explaining exactly this, and for this the concept of structure—and not so much for the question of deciding between subjectivity or objectivity—is the important thing.

A last point regarding the question of why one speaks of self-organization. A system can operate only with self-established structures. It cannot import structures. This too needs to be explained. If you look at research on the acquisition of language, it is nearly impossible to understand how a child can learn language so fast. Some of you will know that Noam Chomsky attempted to solve this problem with the postulate of natural and innate deep structures, which, however, could never be discovered empirically.[3] Presumably modern research on communication would suppose that one learns language in communication itself, as one who is called upon by speakers, if one may say so—speakers who simply suppose that the one who is called upon understands even if they know that he does not yet understand. Through practice there arises the habit to discern specific noises as language and then to repeat specific meanings. This explanation would at least not contradict the thesis that the structure can be established only within the system itself.

If one thinks otherwise, one would have to imagine that someone who learns how to speak is educated in a specific sequence. He does not begin speaking himself; rather, it is dictated to him how he is supposed to speak. Thus, however, one would face the difficulty of explaining the varieties of linguistic development. Research on dyslexia, as well as on writing and reading errors, has shown that the tendency to make mistakes varies greatly within a given class of students. It is impossible to come up with a single didactics of reading or writing because the tendency to make errors differs from child to child. One reacts more on phonetics, another one likes to shorten words and omits a letter, so that the individuality of the process of language learning and then of learning to read and write is extremely high, much higher than commonly supposed. In order to be able to take this factor into account, the shift toward self-organization in the process of learning is, I believe, indispensable. This is not to say that an observer cannot still find out if the words that a child has learned are the same as in the dictionary or as other people use them. But one cannot explain this as an importation of structures, but only on the basis of structural coupling of systems to each other and to their environment.

On the level of abstraction of a general theory, we still have only very little knowledge about how structural development actually takes place. I imagine that, in any case, it does not function like the making of a thing of which you know the necessary components and then put them together. The peculiarity of structural formation seems to consist in that one first has to repeat—that is, one has to recognize any given situation as the repetition of another one. If everything were always completely new, one could never learn anything, and although, of course, everything is always completely new—all of you look different today than during the last lecture, sit at different places, watch differently, sleep differently, or write differently; if one takes a concrete look, every situation is incomparable—there is this phenomenon, which is hard to describe exactly, that, for instance, one recognizes faces. Knowing why it is that you recognize somebody or describing the one whom you recognize is significantly more difficult than the recognition itself. You know the sketches of faces made from eyewitness accounts in newspapers that result from such descriptions, but only with great effort and the help of computers, whereas recognition otherwise normally functions quickly. In order to be at all capable of repetition—and this is once more a circular argument—we have to repeat recognition; that is, we have to be able to do two things: first, identify—that is, classically speaking, recognize essential features or clues of identities—and, second, generalize, in the sense that we can again make use of the identity in spite of the different nature of the situation and sometimes very significant dissimilarities. We are dealing first with a limitation or a condensation of something, and, caused by this, at the same time also once more with a generalization, in the sense that we recognize the same people in completely different contexts, in completely different situations, and often after many years, or we are able to reuse, in language, the same words although we use them in different sentences, on another day, in another mood, in the morning instead of in the evening, etc.

This theory seems to confirm that the continuous testing of identification and generalization—or, to put it even more paradoxically, of specification and generalization—can be the effective property (*Eigenleistung*) only of a system that can be produced only in either the psychic or the communication system. If this communication system were not functioning, we would never learn language. The communication system provides words or standardized gestures that are repeatable and also usable in different contexts, even though there are different effects, since a word denotes highly varying meanings depending on the sentence in which it is spoken. And this ambivalence or this paradox of specification and generalization seems to me to be the reason why this can develop only within a system. Otherwise one would have to imagine that the production of something could be learned like the making of something following a recipe and, of course, by following directions.

To bring this part to a conclusion: Maturana has spoken of "structurally determined systems," and this expression in a certain fashion has found its way into the literature.[4] But if one takes the expression literally, this is only one-half of the matter. The operations of a system presuppose structures; otherwise, one would have only a limited repertoire with a set of fixed structures. The larger the number of structures, the greater the diversity, and the more recognizable is the system for itself as the determining factor of its own state and its own operations. On the other hand, the opposite is also true. The structures in turn can be formed only by its own operations. It is a circular process: The structures can be formed only by its own operations because its own structures in turn determine the operations. This is obvious for the biochemical structure of the cell since the operations serve at the same time the purpose of the formation of the programs, here the enzymes that are used by the cell to regenerate both the structures and the operations. In the social system, if one thinks of language, the same is the case. Language is possible only on the basis of the operation of speech; one would soon forget one's language if one could never speak and had no opportunity to communicate—or rather not even learn it in the first place. Conversely, language is the condition of speaking. This circular relation presupposes as framing, as condition, the identity of specific systems within which this circle is brought into operation or transformed into sequences, such that time can dissolve the circle. That is, the relation is circular only if abstracted from time. In reality, however, there are operations that establish with a minimum of structural effort more complex structures, which, in turn, enable more differentiated operations to take place.

By the way, this is also the point of contact where considerations regarding the theory of evolution—which I left out in this lecture series—enter the discussion because it would be the task of the theory of evolution to explain why single inventions such as the biochemistry of life or meaningful communication, the meaningful exchange of signs, develop into highly complex systems or a great variety of species—that is, how an abundance of structures can be formed although there is, respectively, only *one* type of operation and although the whole thing is set up in a circular way. The structures depend on operations because the operations depend on structures. The whole thing, however, has a flavor of paradox only because the paradox is formulated in abstraction from time, whereas reality makes use of time and can thus develop and so unfold the paradox.

The second part of this section regards the concept called "autopoiesis." The starting point of the considerations is that what has been said about structures is also valid for the operations themselves. I have already touched on this since otherwise I would have been unable to depict the concept of structures that are circularly intertwined with operations. Conversely, the theory of autopoiesis—I will soon get back to the meaning of this expression—is the condition of the possibility of depicting structures in the way I have just attempted. Autopoiesis means, in Maturana's definition, that a system can generate its own operations only through the network of its own operations.[5] And the network of its own operations is in turn generated through these operations. In a certain way, this formulation contains too much of a proposition, and therefore I have attempted to pull it apart. On the one hand, we are dealing with the thesis of operational closure. The system generates itself. It not only produces its own structures, such as certain computers can develop programs for themselves, but it is also autonomous on the level of operations. It cannot import any operation from the environment—no foreign thought enters my head, if it is taken seriously as a thought—and no chemical process can become communicative. If I spill ink on my paper, it becomes illegible, but this does not produce a new text. This operational closure is only a different formulation of the proposition that an autopoietic system generates the operations that it needs to generate operations through the network of its own operations. Maturana, in his Chilean-American English, speaks of "components." This concept lacks clarity insofar as—but also covers a lot because—it leaves open the question whether "components" are to be understood as the operations or as the structures. For the biologist this distinction may be less important since he is not dealing only with this fixation on event-like operations, but also with chemical states and with the alterations of chemical states within the cell; thus he can still try to coin the elementary concept as a state, albeit one of short duration. However, in the areas of the theory of consciousness or the theory of communication, the event-character of elements that cannot be further dissolved forces itself upon us. A sentence is a sentence; it is spoken when it is spoken and no longer afterwards and not yet before. A thought or a perception, when I see something, is current in this moment and no longer afterwards and not yet before, so that the event-character of the operations becomes obvious. It is my feeling that this leads to a stricter distinction between structures and operations and to the dispensation of the overarching concept of the "component." But this remark is intended only as a reading direction for texts by Maturana. On principle, I do not see a decisive difference.

But why "autopoiesis?" Maturana once told me how this expression came to his mind. Initially he had worked with circular structures, with the concept of a circular reproduction of the cell. The word "circular" is a common one that does not create further terminological problems, but for Maturana it lacked precision. Then a philosopher, on the occasion of a dinner or some other social event, gave him a little private lecture on Aristotle. The philosopher explained to him the difference between "praxis" and "poiesis." "Praxis" is an action that includes its purpose in itself as an action. Aristotle here meant the ethos of life in the polis, its virtue and excellence, called "arête," whose importance is not due to its contribution toward the creation of a good city but rather already makes sense on its own. Other examples would be swimming—one does not do it in order to get somewhere—or smoking, chatting, or the reflections in universities, which too are actions satisfying as such without leading to any results. The very concept of "praxis" already includes self-reference. "Poiesis" was explained to Maturana as something that produces something external to itself—namely, a product. "Poiesis" also implies action; one acts, however, not because the action itself is fun or virtuous, but because one wants to produce something. Maturana then found the bridge between the two concepts and spoke of "autopoiesis," of a poiesis as its own product—and he intentionally emphasized the notion of a "product." "Autopraxis," on the other hand, would be a pointless expression because it would only repeat what is already meant by "praxis." No, what is meant here is a system that is its own product. The operation is the condition for the production of operations.

It is also interesting that the concept of "poiesis," of the making, of the manufacturing, as well as even more obviously the concept of "production," never implies the total control over all causes. One can always control only a partial segment of causality. If, for instance, you want to cook an egg, then you think of the need for an electric stove or that you somehow have to make a fire, but you do not think of having the ability to change the air pressure or to modify the composition of the egg in such a way that it would cook by itself. I do not know if such things would be technically possible, but there are a large number of possible, normally presupposed causes in processes of production that could be varied in order to come to new methods of production. The concept of production contains the idea that one is not dealing, in the classical language, with "creatio," with the creation of everything that is necessary, but only with production—that is, with the bringing forth out of a context of preexisting conditions that can be presupposed. This is not unimportant because in the discussion of autopoiesis it is stated time and again that humans, for instance, are indispensable causes of communication. Then, however, one could name other things as well: blood circulation, a moderate temperature, the normal electromagnetism of the earth so that bone fractures heal again, and other environmental conditions for communication. Neither the concept of operational closure nor the concept of autopoiesis denies this. Such a denial would also not fit into systems theory, which always proceeds from a difference between system and environment.

We are thus dealing with "poiesis" in a Greek or traditional and strict sense of production, the manufacturing of a product, combined with "auto," which is to say that the system is its own product. It is not only a self-evident praxis. In a certain way, some effects of the uncommon character of this expression, or of the conspicuous nature of this word that had not previously been known, are that the theory of autopoiesis is both overestimated and underestimated and that there is a lot of criticism that is not to the point. On the one hand, there is the criticism that it is a biological theory that should not be transferred to other realms. This is understandable since in the realm of biology the infrastructure, if one wants to say so, or the chemistry of autopoiesis is clarified. Normal biologists therefore ask what they gain in addition to what they already know when they now also call this "autopoiesis." The biochemistry of the cell is known, so why then the expression "autopoiesis"? In fact, the concept made it easier to search for answers to the questions: What is operational closure? and What is the difference between production and causality? Still, it is a coincidence that it was biologists and neurobiologists, working on a terrain that had been prepared in a certain way, who invented this concept. Maturana invented it. Varela took it over. But this does not mean that its usage in other realms is merely, in a strictly technical sense, an analogy. An analogy is founded either on an ontological proposition that the world has an essential structure that produces similarities everywhere since it is meant this way by creation, or on the argument that because it is like this in the living realm, it also has to be like this in the psychological or social realm. But one does not need this argument. If one defines the concept in a sufficiently abstract way, it becomes clear that it can also be used in other cases, if one succeeds in demonstrating this.

I have had relatively long discussions with Maturana on this issue, and he always said that if one talked about the autopoiesis of communication, then one would have to be able to demonstrate it. One thus has to demonstrate that the concept indeed functions well in the realm of communication, so that one can say that any particular communicational act is possible only within the network of communication. It cannot be conceived of as an event that happens only once. It can also not be produced from the outside, in a communication-free context—as a chemical artifact, for instance—and then have communicative effects; instead, it always has to be produced by communication. I think that this does not create great difficulties. One can see quite easily—particularly if one considers the linguistic tradition, Saussure, for instance, and what grew out of this—that communication proceeds along with its own differences and that it has nothing to do with chemical or physical phenomena.

Opposition arises only when Maturana refuses to call communication systems social systems. There is a strong emotional argument on his side. He does not want to leave humans aside and lacks the flexibility with respect to sociological or linguistic issues that would allow him to see how humans can again be taken into account. He does not want to give up the denoting of concrete human beings—who form groups and such things—with the expression "social systems." Only there lies the difference.

In the sociological literature that refuses to integrate this concept into the theory of social systems, there exists the conception that it is a biological metaphor, similar to the metaphor of the organism that is applied in an uncontrolled manner and perhaps with conservative intentions with respect to social systems. Sociologists are sensitive to this point. However, this is a discussion that will phase out in time, I believe. It is already an indication of little concern if someone states that something is a metaphor. If one goes back to Aristotle's *Politics* and other traditional texts, one can say that all concepts are metaphors. Everything is somehow generated metaphorically and it then becomes independent, technically, so to speak, when applied in language with its methods for condensing, identifying, and enriching usage possibilities. If one has this meaning of "metaphorical" in mind, then nothing can be said against a metaphor. But this would then have to be generalized so that one had to say, for instance, that the concept of "process" is metaphorical too. It comes into sociology from philosophy, into philosophy from juridical terminology, into juridical terminology from chemistry, or vice versa; I cannot trace this back so exactly. The point is that in the end, everything is metaphorical.

Another side of the discussion is more important. I also think that the concept of autopoiesis and the theory of autopoietic systems are both underestimated and overestimated. The radicalism of this approach is underestimated. The radicalism goes back to the thesis of operational closure. The thesis of operational closure implies a radical shift in epistemology, as well as in the ontology it presupposes. If one has accepted this and relates the concept of autopoiesis to it—that is, if one treats it as a further formulation of the thesis of operational closure—then it is clear that it is connected to a break with the epistemology of the ontological tradition that supposed that something from the environment enters into the one who cognizes and that the environment is represented, mirrored, imitated, or simulated within a cognizing system. In this respect the radicalism of this innovation is hard to underestimate.

On the other hand, its explanatory value is exceptionally small. That has to be particularly stressed in the context of sociology. Basically, one can explain nothing with autopoiesis. With this concept, one gains a different starting point for more concrete analyses, for further theses, or for a more complex application of supplementary concepts. Already in biology it holds true that the difference between worms, humans, birds, and fish as a result of the one-time invention of biochemistry cannot be explained on the basis of autopoiesis. The same is the case with communication. Communication is a continually ongoing fact, a continually ongoing operation that reproduces itself. It is not just a one-time event, as when an animal makes a gesture to which other animals react by running wildly, crisscrossing, and then, at some other time, for them, a new orientation and imitation comes into being. When this state of one-time events is overcome, communication, by means of separate signals or signs, presupposes recourse to earlier sign usage and anticipation of possibilities of connectivity. Once this is secured, society is thereby constituted, but it still can be Hottentots, Zapotecs, Americans, or any other culture. That this can vary in time cannot be inferred, with respect to its structural development, from the concept of autopoiesis. This is to say that there is not really a great explanatory value. And this creates problems for a methodologically conscientious sociologist: if one considers that theories are supposed to provide instructions for empirical research and thus structural prognoses, then to recognize theses without an important explanatory value, without hypothesis formation, without the initiation of an empirical apparatus as fundamental, runs against the grain of normal scientific teaching. In line with this consideration, the theory of autopoiesis is a meta-theory, an approach that answers what-questions in a peculiar fashion: "What is life?" "What is consciousness?" or "What is the social?"—"What is a social system independent of its particular features?" The concept of autopoiesis answers these what-questions, and this too is a thought of Maturana's.

At the end of a long course on the evolution of life a student approached Maturana and said to him that he had understood everything but was still wondering what it actually was that had evolved. For the time being, Maturana had to give it a pass. A biologist normally does not ask the question of what life is. And in sociology the question of what the social is is not one that keeps this discipline busy. In psychology as well, questions in this form that ask what the soul is, what consciousness is, are uncommon. What-questions tend to be tabooed, but the concept of autopoiesis aims at these questions.

We are dealing with, if you will, the refoundation of a theory. But this means at the same time that the following efforts demand many more concepts than simply the word autopoiesis. This concept provides little information for our work. Systems theory has to enrich itself on a more general level in order to work with this concept, to make decisions, and to be able to separate phenomena. The topic of structural coupling will become important in the next lecture because Maturana attempts to explain that which he calls "structural drift" on the basis of different structural couplings: The structural development of a system or of a type of systems depends on the structural coupling with its environment to which it is exposed.

Let me add two further points regarding the current discussion of the concept of autopoiesis. I think it is important that one maintain the rigidity of the concept, that one consequently say: A system is either autopoietic or not autopoietic. It cannot be a little bit autopoietic. Regarding "autos" it is clear that the operations of the system are either produced within the system or that they are, in important respects, provided by the environment—for instance, by an imported program that a computer works with. Here an either-or holds true. With respect to life this is clear: one either lives or is dead. It is only for seconds that doctors are in doubt if one is still alive or already dead. A woman is either pregnant or she is not pregnant, but she cannot be a little bit pregnant. This example goes back to Maturana himself, *ipsissima verba*. This means that the concept of autopoiesis is not a concept that can be gradualized. And this in turn means that the evolution of complex systems cannot be explained with the concept of autopoiesis. If one still tries this, one gets to theories that say that a system slowly becomes more autopoietic. Initially it would be totally dependent on its environment and gradually it would gain autonomy, and then, first, structures would become less dependent on the environment, a little more or less, and then the system would become more autopoietic in its operations as well. There is such a tendency. Gunther Teubner in Florence made similar suggestions in order to be able to integrate considerations of evolutionary biology into the theory of autopoietic systems.[6] In the meantime, I also became acquainted with literature in the discipline of economics that gradualizes in a certain way the autopoiesis or the autonomy of companies and arrives at such a concept as "relative autonomy."[7] It is maintained that a system would be relatively autonomous and that in certain respects it would be independent from its environment and in other respects dependent on its environment. In a strict understanding of the word autopoiesis, however, it says nothing about a system's dependence on or independence from an environment since it is a causal question that asks in which respects a system is dependent on the environment or how the environment affects the system. And this is, again, a question that has to be addressed to an observer of the system.

In addition, one cannot assume as a principle the constancy of the sum that would state that a system is the more independent from the environment the less dependent it is. Many experiences indicate that highly complex systems that are autonomous to a high degree—if one can relativize this word—increase in independence and specific dependence at the same time. In modern society, an economic system, a legal system, or a political system is independent to a high degree, but it is to an equally high degree dependent on the environment. When the economy does not flourish, political difficulties arise, and if politics are unable to provide certain guarantees—for instance by means of law—or if politics interferes too strongly, then this creates problems in the economy. We have to distinguish—and this leads again back to the thesis of operational closure—between causal dependence/independence, on the one hand, and self-generated operations on the other.

I am not even so sure myself if this is convincing in the end. We all have, specifically in European culture, a strong conceptual tendency to convert every-thing into causality and then, for instance, to always understand such terms as "an operation generates an operation" or "production" causally. This makes it difficult to conceive of operational closure as completely separated from causal theories. We always glide back to the conception that the thesis of operational closure is a specific thesis about the internal causality of systems. In a certain sense this is even the case. It can be converted to causality. But on principle the matter should be understood in such a way that the condition of connectivity is not a condition that suffices to cause the following state. Exactly this is the theme of the conceptual field of structural coupling.

Notes

1. See von Foerster, "What Is Memory?"
2. Thus Berger, "Autopoiesis."
3. See Chomsky, *Aspects of the Theory of Syntax.*
4. See Maturana and Varela, *Autopoiesis and Cognition.*
5. See the definition in Maturana, "Autopoiesis."
6. See Teubner, *Recht als autopoietisches System,* 38 ff.; see also the printed version of the discussion in Teubner and Febbrajo, *State, Law, and Economy.*
7. See Kirsch and zu Knyphausen, "Unternehmen als 'autopoietische' Systeme?"

Space Is the Place

The Laws of Form *and Social Systems*

MICHAEL SCHILTZ

The single most striking characteristic of George Spencer-Brown's *Laws of Form* is the variety of misunderstandings concerning its reception.[1] Its basic idea is actually quite easy: "form" or "something" is identical to the difference it makes (with anything else) and (thus) eventually different from itself. All "something" or "form" or "being" is explained as the residual of a more fundamental level of *operations* (namely, the construction of difference), including the "calculus of indications" explaining the very *Laws of Form*. Due to its constructivist nature, the calculus has enjoyed admiration from a variety of people, some of whom are regarded of major importance in their respective scientific disciplines. After a meeting with Spencer-Brown in 1965, the philosopher and logician Bertrand Russell congratulated the young and unknown mathematician for the power and simplicity of this calculus with its extraordinary notation. In 1969, shortly after the publication of *LoF*'s first edition, the father of neocybernetics, Heinz von Foerster, enthusiastically described it as a book that "should be in the hands of all young people."[2] In the cybernetic tradition, by the way, *LoF*'s resonance is undiminished. The international journal *Cybernetics and Human Knowing* published a Charles Sanders Peirce and George Spencer-Brown double issue in 2001; there exist two extensive Web sites with *LoF* material and new Spencer-Brown mathematical work (see "Spencer-Brown–related sources" in the notes below); and a revised English edition of *LoF* is forthcoming. One would conclude that *LoF* is very much alive indeed. But as noted above, appraisal for the calculus is certainly not univocal. There exist (some very advanced) criticisms of the calculus. Some authors regard it as misconstrued from its very beginning: for Cull and Frank, the *Laws of Form* is no more than the *Flaws of Form*. The greater bulk of disapproving comments is, however, less than a spelled-out, intricate argument. In general, it aims at the status of *LoF* within the mathematical tradition and rejects it as a mere variant of Boolean algebra, simply using a new notation. *Nil novum sub sole*, so to speak. Whatever be the case, *LoF*'s thinking, especially where it concerns its far-reaching constructivist implications, has clearly not yet been well established. Spencer-Brown's (promising) claims notwithstanding, the context of his work, its notation, and its exotic vocabulary need a great deal of clarification.

For that very reason, the adoption of the calculus in contemporary sociological theory cannot be an obvious course. And yet some sociologists—most notably sociologists working in the systems-theoretical tradition of Niklas Luhmann—work with it "as if it were not only common knowledge, but as if one had fully grasped the transformation of the deep ontological structure it induces."[3] This is most certainly a reason for surprise or doubt. What does it mean when concepts (as "forms") are consistently formulated as a distinction? In an example, what is the sense in defining systems theory as the theory of the difference between system and environment? Next, is Luhmann's use of paradox, another central notion in *LoF*, more than an inflated postmodernist rhetorical device? Why does Luhmann insist on constructing a circular epistemology (that is, sociology as a way of society to picture itself in itself)? And why, above all, does Luhmann claim his theory to be universalist *yet not* solipsist? It is deplorable that the aforementioned reasonable doubts have generated a stream of publications harshly rejecting all Luhmannian theory. Danilo Zolo, for instance, has denounced the theory as a very complicated version of circular reasoning. Gerhard Wagner, on the other hand, specifically attacks Luhmann's epistemological grounding in *LoF*. Those differentialist claims, so Wagner argues, are no more than the foundationalist or essentialist thinking to which Luhmann himself claims to react.[4]

In the following, I intend to tackle exactly these problems. I will do so by systematically discussing the construction and argumentation of *LoF* (1) as I believe it holds the key to itself and to the sociological claims of Niklas Luhmann, but also (2) because such analysis has been conspicuously absent in the existent literature. A great deal of attention will be paid to problems associated with the circular construction of the calculus. At all times one should be aware of the difficulties or impossibilities of presenting circularity in a circular way; at least the medium of the book or oral presentation demands that we proceed linearly—that is, from the first to the last page respectively, or from the opening to the concluding remarks. As we shall see, this limitation (or, if one wants, "paradox") contains the solution to a rightful understanding of *LoF* and Luhmannian theory construction: "Reason, or the ratio of all we have already known, is not the same that it shall be when we know more" (William Blake). I will start by showing how *LoF* relates to the mathematical tradition and how this refutes a great part of existing criticism. Next, after a treatment of the calculus's topological notation, I will briefly show the linear development of the calculus out of the primary arithmetic and primary algebra. Especially the planar foundation of both allows us to understand why the grounding of a theory of (social) systems in Spencer-Brown's calculus can hardly be an obvious course of theory building. Yet the presentation of self-reference in the calculus notation, as Spencer-Brown demonstrates, is possible if and only if we are prepared to change the medium in which we are writing. Self-reference defies presentation in plane space yet can be presented in topologically more intricate versions of space. As we will see, the latter corrodes our most profound ontological presuppositions radically; on a higher level, it is also responsible for sharp differences between the systems theories of Talcott Parsons and Niklas Luhmann. Last but not least, *LoF*'s altered treatment of space (and its relationship with time) allows an exploration of the peculiar epistemologies of both Luhmann and Spencer-Brown.

Starting Point: A Nonnumerical Arithmetic

Before the discussion can commence, I should draw the reader's attention to the extraordinary economy with which Spencer-Brown equipped the calculus. This is *not* a common logical calculus founded on postulates. It is not even a logical calculus. *LoF* must be studied as a book of mathematics, an "arithmetic whose geometry as yet has no numerical measure." No numerical measure, indeed! What does that mean? As Spencer-Brown rightfully underscores, we find ourselves here in the primitive a priori dimension of all notation: two-dimensional space.[5] What we will be doing here, to put it bluntly, is drawing figures in sand or on a piece of paper. Interestingly, this is what constitutes the challenge for a twentieth- or twenty-first-century public: the level of investigation is as deep as being "beyond the point of simplicity where language ceases to act normally as a currency for communication."[6] Our investigations will have to take place at a prediscursive level, where "something" comes into being so to speak.[7] This is very different from the level of number, and certainly logic, which is not so much concerned with the world but with the rather limited domain of our (human) *cognitive relationship with* the world. "Logic is not, and has never been, a fundamental discipline," Spencer-Brown therefore argues.[8] And for the same reason, postulates cannot exist here. Spencer-Brown departs from a very basic experience of dealing with the world, with "things," "stuff." This in itself makes the calculus a true rarity in the history of mathematical thought. I concur with von Kibéd and Matzka that Wittgenstein's *Logisch-philosophische*

Abhandlung (the *Tractatus*), and in several respects Charles Sanders Peirce's work, are the kind of inquiry closest to the one presented in *LoF*.[9] In view of its rigorous confinement to the very fundamentals of logical systems, *LoF* is most acutely referred to as a protologic: a research into ordinary arithmetic, rather than ordinary algebra, or "an inquiry into the pre-discursive laws emerging with the most elementary position of 'something.' These laws must be situated at a level preceding the level of expression grasped by classical logic."[10] Thus, *LoF* anticipates and wards off the major part of its critics at its most elementary level. This is important: when Kuroki Gen, a harsh critic of *LoF*, describes the work as a reformulation of propositional logic or Boolean algebra, he is at least neglecting the calculus's construction and is possibly ignorant of the meaning of its very starting point.[11] We will encounter the consequences of this misunderstanding below.

This being said, we can proceed to what Spencer-Brown grants us in order to commence calculating (chapter 1 in *LoF*). And that is not much. Spencer-Brown is very cautious not to break with the objective of starting from the very beginning. He simply delivers a definition of form ("Distinction is perfect continence") and two axioms contained in the definition:[12] (1) The law of calling refers to the descriptive aspect of distinctions. Once (a delineated) something has been given a name (call), recalling it does not alter it—"The value of a call made again is the value of the call." (2) The law of crossing concerns the injunctive or more clearly operational aspect of distinctions. Here a difference "does make a difference." One can only be in the form, or not—"The value of a crossing made again is not the value of the crossing."

Yet its mathematical economy notwithstanding, let us not be mistaken about the *LoF*'s intentions: "The theme of this book is that a universe comes into being when a space is severed or taken apart. The skin of a living organism cuts off an outside from an inside. So does the circumference of a circle in a plane. By tracing the way we represent such a severance, we can begin to reconstruct, with an accuracy and coverage that appears almost uncanny, the basic forms underlying linguistic, mathematical, physical and biological science, and can begin to see how the familiar laws of our own experience follow inexorably from the original act of severance."[13] As will be clear, this passage contains *LoF*'s undeniably universalistic (and thus circular) aspirations: starting out from an original act of distinguishing, *LoF* intends to describe its consequences for (1) the possibility of the world ("things" as form), (2) the possibility of developing a (cognitive) relationship with the world of things (knowledge or "cognitive categories" as form), and (3) eventually, the possibility of describing the possibility of discovering these possibilities (the laws of form as the precondition of all form, or the universe as "what would appear if it could").[14] The last concerns the pure circularity of the calculus: the Form as an explanation of itself. It is this part as well that led Heinz von Foerster to link *LoF* with Wittgenstein's "problem of the world"—that is, the fact that the world we know is constructed in order to see itself, while that appears to be a logical impossibility.[15]

Forms Taken out of the Form

But it is too soon to discuss the link with Wittgenstein. Let us commence calculating. As should be expected from this basic inquiry, the calculus—which starts in chapter 2, subsequent to the outlining of the conception of the primal form— begins with a command of surprising naïveté: "Draw a distinction!"[16] "Draw a line," "make a distinction," is the primal injunction. As such, Luhmann would say, one performs the operation of "observation." One de-lineates something and *simultaneously* indicates one of the sides separated by the distinction.

In order to express the conception of the "form" through a formal notation, Spencer-Brown employs the ¬ , the "mark of distinction," a topological notation. At this stage of the calculus, the mark ¬ represents a "cross" (descriptive) that also ought to be taken to mean "cross!" (injunctive). The mark is, in Peirce's sense, a *portmanteau-symbol*: it combines both the aspects of plain denotation and an injunctive or instructive meaning to cross the distinction and indicate one of the separated sides.[17] "Let any token be intended as an instruction to cross the boundary of the first distinction. Let the crossing be from the state indicated on the inside of the token. Let the crossing be to the state indicated by the token."[18] This is no more than stating the obvious. When we draw a distinction (for example, a circle), then the distinction cannot be neglected; it has affected the space in which it is written, and we are, as such, "in" the form. The first distinction literally is a first judgment, an *Ur-teil*, which determines everything coming after it. Once the distinction has been drawn, a "universe" is there, and the gates to return to a state of nothingness are closed; that world is the mere "nameless origin of heaven and earth," the phenomenology of which is lost.[19] The notation ¬ (alternatively ( ) or <<>>) thereby expresses that topological *asymmetry* as well. Simultaneously with the drawing of a distinction, one of the sides is indicated. The concave side of the mark thereby represents the "inside" of the form or the "marked state." The other side is the outside; it is a nameless residual, an unmarked leftover, from which the marked side is delineated. We must stress here that the drawing of the distinction and the indication of one of the separated sides are two simultaneous aspects of one operation. In Spencer-Brown's terms: "We take as given the idea of distinction and the idea of indication, and that we cannot make an indication without drawing a distinction."[20] It is equally impossible to draw a distinction without making an indication. Why else would one draw the distinction at all? The indication is the motive of the distinction. For the time being, we can thus conclude that form, indication, and distinction are *implied* within each other, not to say identical. For it is clear that every observation implies the drawing of a distinction, and this implies in turn that every form has to be conceived of as a distinction: "Call the form of the first distinction the form."[21] This then allows the reader to follow what could be called a syntax of form (*LoF*, chapter 2): aspects of the form ("name," "content," "value," etc.); basic calculatory possibilities ("the form of cancellation"/"the form of condensation"); and more complex notions—for example, the "unwritten cross."

In a next step (chapter 3 and further), the calculus is developed linearly (!) out of the aforementioned syntactical complex—the latter represent the rules of the game, so to speak. It grows into a *primary arithmetic*, which serves, in turn, as the foundation for a *primary algebra* (chapters 4 through 6; the line of the argument does not demand we enter theorems consequences or canons developed here).[22] At several points in the appendices, but also in the preface(s), introduction, and the "note on the mathematical approach," the difference between these is stressed, and Spencer-Brown definitely favors the primary arithmetic over the more commonly investigated level of algebras—that is, Boolean algebra.[23] Arithmetic, he says, is about the constants, the individuality of form, and the individuality of the calculatory relationships the form builds. The algebraist, on the other hand, "is not interested in the individuality of numbers, he is interested in the generality of numbers. He is more interested in the sociology of numbers."[24] The formulation of the arithmetic was thus formative to the development of an algebra (Boolean) in the first place: "So, to find the arithmetic of the algebra of logic, as it is called, is to find the constant of which the algebra is an exposition of the variables—no more, no less. Not just to find the constants, because that would be, in terms of arithmetic of numbers, only to find the number. But to find out how they combine, and how they relate—and that is the arithmetic."[25]

Elena Esposito has speculated that both the primary arithmetic and the primary algebra may be instrumental to a sound understanding of cybernetic *constructivism*, especially where it relates to the difference between first-order and second-order observations.[26] It all boils down to appreciating the arithmetic and the algebra as autonomous parts of the calculus that correspond to existent (because observable) systemic levels, the level of elements and the level of systemic organization respectively. (1) The arithmetic, as Esposito argues, represents the "formalization of autopoiesis." Here everything is about the execution of rules relating to constants ("specific objects"). It is a first-order level of observation, a level where there can be awareness only of the existence of form. (2) The algebra, which could be constructed through the "theorems of connection," is a different matter. This truly is a *calculus taken out of the calculus*, a level where the forms developed in the arithmetic are object of other forms at a higher level (in turn bound by the rules of the arithmetic!). It is about arrangements (and hence indication *and distinguishing*) of arithmetic forms (*distinctions*). In a sense, it is the systemic level. Esposito explains: "The algebra formalizes a specific type of autopoiesis (and thus requires the validity of all arithmetic formulae): the autopoiesis of a system, the operations of which are observations. Yet, it remains a fact that its operations (including the operations of observation) can only be observed by an external observer. This contains an openness which implies and eventually requires the integration of first order observations with observations of higher orders."[27]

Yet however tempting Esposito's metaphorical use and/or explanation of the different levels may be (and I am certainly not saying Esposito's remarks are entirely wrong!), it is mathematically unsound: autopoiesis is an issue in neither the primary arithmetic nor the primary algebra. What we are doing here is still drawing distinctions in the plane, discovering how they relate, how they may cancel each other out.

Intervention 1: *Laws of Form* and Social Systems

The reader will understand that this leads to further doubts: how is the notion of "form" in the calculus of indications to be linked to a theory of social systems? By no means should the calculus of indications be understood as a "brand" of systems theory. I repeat: the calculus is best viewed as a *protologic*; it was primarily written in reaction to some assumptions held in logic. And being a *proto*-logic, it was not even stamped as a logical calculus, but as a *mathematical inquiry*. Another difference, however, between the Spencer-Brownian calculus and Luhmann's systems theory is that the former mainly represents finite forms (as the calculus demands, they exist as a finite number of crosses), whereas social systems *by definition* hold out the prospect of *infinity*. They have never been set in motion "at a certain point in time," as such a point would presuppose an earlier communication to which they could connect; and *vice versa*, they don't break off at a certain point, as such a point would hold out the prospect of continuation. In Spencer-Brown's calculus, potentially infinite forms are mentioned only where the calculus has been taken "so far as to forget it."[28] An attempted coupling of *LoF* and the theory of social systems should therefore come as a surprise.

Let us take a look at how an apparent contradiction contains the key to its solution. First of all, how do we define a system? A system exists when there is something capable of identifying a specific operation as belonging to itself—that is, when there is something capable of discriminating that operation from operations that do not belong to itself (with attention to the sheer tautology of systemic operations). A system then uses the products of preceding systemic operations for the performance of new and different operations, again identified as belonging to the system, not to its environment. Systems thus necessarily carry an image of what they are not, although in a truly ambiguous way. This operational mode is designated as self-reference. Self-reference expresses the unity the system creates for itself. It indicates that a system refers to itself in all its operations: "There are systems that can develop a relationship with themselves and can distinguish this relationship from relationships with the environment."[29]

This conceptualization of self-reference in the terms of the *LoF* does not seem a self-explanatory course. After all, it would come to mean that forms, distinctions (for example, system/environment), develop a relationship with themselves, although self-contact is implied as being an impossibility in the definition of the "primal form." Once more, I stress that the operation(s) of distinguishing and indicating should be studied as a single operation. And once more, I must emphasize that the indication is the one and only motive of the distinction. Taken together, does this not mean that the distinction is employed simply in order to be forgotten in the indication?[30] Consequently, if the whole range of the distinction is in itself the residual of an observation, how can self-reference possibly be realized? Is not self-reference an impossibility, as it implies the distinction's capability of referring—that is, indicating—itself in itself and employing earlier indications for the production of new indications in the (same) form? Is self-reference not excluded in the very definition of the primal form? In brief: is self-reference not inhibited because of the fact that distinctions, differences, "make a difference"?

Reentry of the Form into the Form

And yet: is this really true? Do all distinctions make a difference? In some obviously neglected but crucial passages of the "appendix" to the calculus, Spencer-Brown reminds us of the use of covert conventions in mathematics: we have agreed to some rules without being consciously aware of the fact we did so.[31] In the beginning of the calculus, and for that sake in this essay, the reader has, for instance, assumed that the distinctions were drawn in a *plane*: a piece of paper, for instance, or the surface of the earth. As we know, distinctions drawn in a plane do indeed build a distinction. But the use of a different mathematical medium makes things a lot more complicated, to the point that seemingly obvious facts are, in fact, not self-evident at all. If, at the outset, we had confined ourselves to writing on a torus (a "doughnut"), for instance, the distinction *would not have constituted a distinction*.[32]

Clearly, our unconscious choice to write in a plane, on a piece of paper—*that* has made the *real* difference. If we are only willing to work with a different medium, with a different conception of the spaces in which the distinctions are drawn, it may very well produce a wholly different arithmetic, algebra, and logic. Such willingness would, moreover, not be without a cause. As we explained above, we do in fact assume the existence of forms (that is, systems) that are thoroughly self-referential, that thus demand a different topological treatment (as they defy representation in the limiting terms of a plane surface or even Euclidean space). During his career as a civil engineer for British Railways, George Spencer-Brown and his mysterious brother, D. J. Spencer-Brown, had already developed special-purpose computer circuits that exhibited all characteristics of self-referential expressions, the prohibition of their use in conventional logic and mathematics notwithstanding.[33] For Spencer-Brown, the question is thus a purely mathematical one. His interests lie with showing the *validity* of imaginary values (for example, $\sqrt{-1}$), the *use* of which has been common in, for example, electromagnetic theory. As they can be used meaningfully for the solution of equations that cannot be solved otherwise, we must accept "imaginary" as a "third" category independent from (1) true (tautology: $x = x$) and (2) untrue (contradiction: $x = -x$). For Luhmann, the problem is to describe self-referentially operating social systems, consisting of operations that take their own results as a base for further operations. These are forms that "in-form" themselves.

In the mathematics of *LoF* (chapter 11), the following solution for our problem is proposed: let us conceive of "space" in a different, operational, way—that is, space as a relation between elements.[34] If we are able to abandon the idea of space as a Euclidean "container" (that is, space as something "in which" things are positioned), it is indeed possible to conceive of self-reference as forms turning up in their own form.[35] Back to the calculus: Spencer-Brown insists that we must therefore allow some way of *contact* between the separated sides of the distinction written in the plane surface. In order to show the self-reference of a form/distinction, the distinction must, quite literally, be *undermined*. Let us therefore dig holes, tunnels, under the surface in which the distinction appears and "corrupt" (from Latin cor-rumpere, to break together) the cross.[36] That space is a torus. If considered operationally, distinctions written on a torus can subvert (turn under) their boundaries, travel through the torus, and reenter the space they distinguish, turning up in their own forms, thus capable of developing some kind of contact with themselves.[37]

Clearly, such self-referential form cannot be decided (from Latin de-cedere, "to cut off") in the plane. The marked state cannot be clearly distinguished from the unmarked state anymore, leading to "indeterminacy." The form is neither marked nor unmarked. It is an imaginary value, flipping between marked and unmarked, thanks to the employment of time.[38] However, this does not preclude its existence: "The value [of self-referential forms] being indeterminate in space, may be called imaginary in relation with the form. Nevertheless . . . it is real in relation with time and can, in relation with itself, become determinate in space, and thus real in the form."[39] Self-referentially operating systems should thus be understood as the operational difference between themselves and their environment, a difference that is made through some sort of self-referential oscillating between the two sides of the distinction (that is, system and environment). By means of self-reference, the environment "out there" can be observed as being drawn topologically into the "inside" of the system (compare the inside and outside of a Möbius ring). This is the meaning of reentry: the two sides of the distinction are reinserted into one of its parts.

Spencer-Brown repeats the notational ramifications of such subversion: "In a simple subverted expression of this kind, neither of the . . . parts are, strictly speaking, crosses, since they represent, in a sense, the same boundary. It is convenient, nevertheless, to refer to them separately, and for this purpose we call each separate . . . part of any expression a marker. Thus a cross is a marker, but a marker need not be a cross."[40] The distinction could thus also be said to have been alienated from its original intent or motive (that is, indicating one separated side), and this by value of being processed within the form (system), in order to safeguard the difference between itself and other distinctions over time. The aforementioned notational arrangement does, however, have intriguing consequences for the form's being. The excursion through the tunnel of the torus and the consequent time employed to return into itself make the self-referential form peculiarly look as shifting between what it is indicating ("cross!") and what it uses to make indications ("marker"). The self-referential form is flippety: "I am the link between myself and observing myself" (Heinz von Foerster). In the parlance of ontology (cf. infra): the self-referential form is both identical to and different from itself.[41]

Laws of Form as Ontology

The mathematical visualization of self-reference in mind, it may be instructive to reconsider an important critique on Luhmann's notion of "system." Interestingly, some authors (most prominently represented by Gerhard Wagner) seemingly assume that Luhmann has proposed to simply replace a thing's or system's *identity* by means of a difference—namely, the difference between system and environment—just like that. Obviously, such a shift has not been the case. Rather, what is lacking *in the critique* is due attention to the growth of the calculus's injunctions in the direction of the "form" of self-reference. Whereas Luhmann himself has, on several occasions, referred to systems theory as an invitation to draw a distinction between system and environment, that distinction is an obvious advance on the topological ramifications prominent only in the concluding chapters of the calculus. Clearly, Wagner mistakenly views the difference between system and environment as the *immediate* offspring of the primal construction, "Draw a distinction!"; his argument has not "followed" up to the coda, let alone self-reference having been "understood." He is still *in the plane.* Consequently, he is myopic concerning the operational aspect of self-reference. Hence, the difference between system and environment cannot be evaluated in its full complexity. As a matter of fact, it must ultimately come to be seen as a rather trivial reiteration of foundationalism: "The fact is that, by the way in which Luhmann understands foundational difference, he practically commits his position to identity."[42] In such a mistaken view system and environment can easily and erroneously come to be conceived of in terms of a polar opposition, quite similar to Hegel's notion of negativity.[43] As my remarks have hopefully made clear, the elaborate notion of a system as a self-referential form can be realized only in the more advanced chapters of *LoF*—namely, chapters 11 and 12—in which there is made mention of the *reentry.*

The system's self-reference can thus only be defined as the act of self-reference, as self-referential performance. And this, as we have seen, demands a quite intricate topological arrangement. The system must reflect, in the formal sense, itself and its environment as a corollary of itself *in all of its operations.* It can secure the connectivity of its operations only by establishing itself as an imaginary value and by employing the time of the tunnel to develop a relation with itself. In short, imaginary space is the only topological possibility for a system to be systemic. For that reason, the difference between the system and the environment cannot be an essentialist difference, let alone a new version of foundationalist thought: "It does not cut all of reality into two parts: here system, there environment. Its either/or is not an absolute; it pertains only in relation to the system, though objectively. It is correlative to the operation of observation, which introduces this distinction (as well as others) into reality."[44] This operational aspect apparently provides the clue to the numerous misunderstandings about Luhmann's ontological premises. And time after time, the peculiar position of the environment has been at the core of the problem. Yet if you are aware of the topological qualities of the torus, you can easily understand that position. The environment is not so much *out there* as *in there*: it simply emerges out of the reentry of the distinction into itself. The environment is *constructed* by the system; it exists only with the form of the system—that is, if there is a boundary that can be employed in order to reenter the system's own inner space.[45]

This form-centered conceptualization has clearly parted with more established distinctions such as *subject* versus *object, man* versus *world*, or *self* versus *difference*. In *LoF* and systems theory, everything, all emergent reality, is discussed as the corollary of a construed difference. So what are the ramifications of self-reference for our "ontology of the world"?[46] Central to our discussion here is the connection between *operationalism* and form. Systemic operations, we have stressed, presuppose self-contact, and, vice versa, self-contact implies systemic "in-formation." Clearly, the form is on its own; it is a self-sufficient and self-engendering reality. Actually, the definition of the primal form, stated at the very outset of the calculus of indications, had already made this clear: "Distinction is perfect continence."[47] But at this stage of the calculus, what we have already known is not the same as what we know now (cf. the citation of William Blake in the introduction). At this point we realize that "form" is the symbol of the world or the universe; *all* form is part and product of a self-engendering self-referential whole,[48] of which even the first form must be embedded in a further form (most fundamentally the difference between drawing a distinction and deciding not to do so).[49] This formal introversion (from Latin *intro-vertere*, turning inside), this very self-reference, refutes essentialism. After all, while we may take it that the universe undoubtedly is itself—that is, indistinct from itself—we must accept the fact that, as self-reference, it is indeed false, or distinct, to itself: "We may take it that the world undoubtedly is itself (i.e. is indistinct from itself), but, in any attempt to see itself as an object, it must, equally undoubtedly, act so as to make itself distinct from, and therefore false to, itself. In this condition it will always partially elude itself."[50] The world is not "what there is."[51] Yet this foundationalist crisis à la Kurt Gödel's theorem of incompleteness should not be seen as a reason for despair. Self-referential paradox, meaning indeterminacy, must be construed as the price systems and the world pay for the possibility of operations, activity, and systemic evolution.

For contemporary systems theory, paradox is not seen as an accident to be avoided, but rather as the creative presupposition of the whole construction. Paradox "is not the fatal end, the definitive failure of all ontological constitution. On the contrary . . . it is the starting point of a history, a movement of system-constitution, full of risks and bifurcations. Paradoxes do not make things impossible, but rather possible."[52]

The French philosopher Jean Clam has therefore speculated that it may be analytically fruitful to employ a difference between *apophantic* and *ergetic* paradoxes, paradoxes occurring in logical expressions (in the plane!) and paradoxes imminent in operations or systemic space respectively. In the apophantic sense, paradoxes do indeed block observation: they do defy determinacy and may therefore be judged destructive or corrupt according to the foundationalist tradition. But the excursion through the tunnel has shown that there is a merit in subversion: what apparently blocked cognition has become an operational loophole, a compelling chance for *system genesis*. Systems must operate in order to achieve the fictitious unity that could not be achieved by the ontologically more elegant way of self-identity and integrity. Systems must operate in order to bypass situations of a structural standstill. And semantically, systems must operate in order to cover up the devastating consequences of manifest inconsistency and contingency: deparadoxicalization. "Operating is always the introduction of a component that avoids the stand-still, because it broadens the space of possibilities. The operating of systems is nothing else than this handling of components which create more possibilities and condensing them into a self-continent but not-finalizable ergetic whole."[53]

Intervention 2: Luhmann versus Parsons

In retrospect, one will agree that the ever-growing prominence of paradox in Luhmann's thinking has changed the concept of social system in ways so fundamental that a sociologist such as Talcott Parsons could not have imagined. At no point, we must concur, can a system be described in terms of invariant structural characteristics. Confronted with the utter impossibility of unity and consistency, in favor of indeterminacy and contingency, systems emerge as mere sequences of ongoing operations. They are no more than a momentary derivative of passing operations, characterized by a self-reinforcing restlessness. Admittedly, Luhmann has never been a committed structuralist. In *Social Systems*, he rejected structuralism on the grounds that "structuralists have never been able to show how a structure can produce an event."[54] His theory of social systems has therefore subordinated structure to function and has shifted the focus from structure to event, the network of which produces the unity of the system, in the event only. But through time, the concept of autopoiesis, which expresses the self-production of the network, has undergone some major changes as well. Whereas the notion had originally been defined in close reference to the way it was designed by Humberto Maturana, Luhmann forsook this definition almost completely in the 1990s. Finally, the notion would be fully rewritten in the terminology of the calculus of indications: "Autopoiesis is thus not to be conceived as the production of a peculiar 'Gestalt.' Crucial is rather the formation of a difference between system and environment."[55] Autopoiesis, the reader will understand, is nothing but the form of a system's basal unrest, the abbreviated expression of the system's concern with getting around its nonidentity. The strong self-referential, and hence reflexive, bias of the notion shows what that means. Enclosing itself in itself—that is, enclosed in itself—the form incessantly crosses its own internal boundary, thereby adding to its level of complexity, but never able to become identical to itself.[56]

This latter point may be helpful in understanding Luhmann's particular brand of functionalist methodology. Already back in 1970, he had criticized Parsonian functionalism on the grounds of assuming a semi-identity of *function* and *causality*.[57] Clearly, the notion of causality, implying necessity and absoluteness, is at odds with a theory that converges around contingency—in politics, in law, in science, in intimacy, in art—in brief, modernity![58] Luhmann therefore returns to Kingsley Davis's critique of functional method and manages to turn this critique inside out: the rejection of functionalism is—in a typically functionalist guise—employed as a solution for some conceptual deficiencies of the functionalist method. First, the relationship between function and causality is asymmetrized: causality must be classified as one exceptional instance of function. Second, functionalism is outlined as a method for *comparing* the potential of systemic arrangements aimed at the maintenance of the system's unity, rather than for indicating the "systematic" relationships between function and achievement. Luhmannian functionalism is a *functionalism of difference* and as such is more than a mere rhetorical upgrading or fine-tuning of a well-known functionalist repertoire. After all, the quest for historically contingent and factually variable functional equivalencies effectively avoids the structural determination of theoretical judgment.[59] With respect to content, attention shifts from the functional arrangement to what could be called the *construction of problems*. And again we encounter self-reference here. Rejecting a social system's (structural) permanence and subscribing to the idea of systems as forms that react to their self-generated complexity, functionalist method also entails adequate attention to the way in which systemic problem solutions (functional arrangements) "trans-form," expand the system's operational status, and thus *reenter* the system's space at a new level. At that level, the problem to which the functional arrangement initially reacted may have disappeared or may be encountered in a different, possibly more accrued, manner. Such are the methodological consequences of self-conditioning self-reference.

Laws of Form as an Autological Construct

Finally, I promised a clarification of the metaposition of the notion of form far beyond the scope of its heuristical application. After all, after indicating the possibility of self-referential forms ("reentry," chapter 11), *LoF* offers a perspective on the position *of the calculus* ("Re-entry," chapter 12). The calculus, as a part of the universe, must be one possible form, distinguishing the forms it has been describing as forms making a difference. The very calculus of indications has been a "tunnel" through which Spencer-Brown and the reader have traveled to arrive at the form of the first distinction, which is now seen as legitimized, justified by all canons, theorems, demonstrations, and proofs that followed it. The "first distinction" was deliberate and historically *contingent*. Yet all that followed was its *necessary* consequence: "The whole account of our deliberations is an account of how [the first distinction] may appear, in the light of the various states of mind which we put upon ourselves."[60] *For that very reason*, the clarification of the laws governing this universe must be considered a trivial matter: "Coming across it thus again, in the light of what we had to do to render it acceptable, we see that our journey was, in its preconception, unnecessary, although its formal course, once we had set upon it, was inevitable."[61] The paradoxical combination of *contingency* (of the first distinction) and *necessity* (of its consequences) demonstrates in what fundamental respect the epistemology of *LoF* differs from classical epistemologies.[62] The calculus of indications, ultimately a function of itself, has established itself as an imaginary value. It can be continued endlessly, as Spencer-Brown does not fail to indicate.[63] On the one hand, its inclination toward the imaginary makes the calculus correlate with what it seeks to describe: like reality or "world," the calculus is "form" that seeks to get a hold of itself but does not manage to do so. On the other hand, its constructivism obviously implies the loss of a privileged position of (scientific) knowledge. At this point it is clear how *LoF* relates to Wittgenstein's problem of the world (cf. supra). Self-reference has come 360 degrees. It is not merely at the root of any possible universe. It also dominates and determines observations of the universe and eventually observations of the laws of form governing both. This should affect scientific observations and scientific method considerably. It implies a shift from a world of things to a world of observations. This is not just a world of the real: "There is a tendency, especially today, to regard existence as the source of reality, and thus a central concept. But as soon as it is formally examined, existence is seen to be highly peripheral, and as such, especially corrupt (in the formal sense) and vulnerable."[64] It is rather a world of the possible and an observer's intention to draw distinctions. Our understanding of the world thus cannot reside in some form of discovery of its present appearance (out there, beyond observation) *but comes from remembering the conventions agreed to in order to bring it about.* The task of the mathematician, whose interest lies with notational elegance and density, may hence lie with bringing the world back to its conventions and abandoning all surplus arrangements. As is well known, Spencer-Brown's conclusions eventually border on the mystical: "To experience the world clearly, we must abandon existence to truth, truth to indication, indication to form, and form to void."[65]

For Niklas Luhmann, however, when presenting a theory of social systems, the challenge is different. His task lies not with abandoning, but rather with expanding. Clearly indebted to *LoF*, Luhmann adopted the notion of form *and the corollary notion of medium.* He has typically used a theory of the latter to identify different types of medium in the social sphere and speculate about their respective topologies, their transformative capabilities, their role in societal evolution, and so on. It is revealing that form/medium came to permeate the whole of Luhmann's theory of society, eventually stretching beyond the original main distinction of system and environment, bringing about obvious problems of theory construction.[66] Epistemologically, furthermore, Spencer-Brown's mathematical conclusions on reentry are expressed as the autology of the distinction between form and medium: form/medium is a distinction, thus form.[67] It, too, must reflect the *triviality* of its *necessity. As a self-referentially organized theory, Luhmann's systems theory represents its own boundary, and its limitations.* The acclaimed universality of the theory can therefore never entail solipsism. If properly observed, the laws of form relate (relativize) the theory's universality to the notation employed, thereby outlining the distinction as its own limitation. Conscious of its social formulation in the social sphere of society, the theory of social systems is simply one possible way of presenting society in society (*die Gesellschaft der Gesellschaft*). It is only one possible form in the all-encompassing medium of meaning. This leads to an interesting question. If the medium of meaning is indeed the ultimate medium of psychic and social systems—that is, if meaning is "the medium of itself"—then what is its "form," the distinction through which it can be expressed? I perceive only one answer: the medium of meaning must be identical to the difference between form and

medium and the reentry of that distinction into itself. Its consequent indecidability is the symbol of our dealing with the world. It expresses the fact that all
our attempts to get a hold of the world are doomed to frustration.[68] Meaning as
our *phenomenology* of this world can only be partial, as the difference between
form/medium can only be actualized as a form. In mathematical terms: meaning
is a lambda-domain occupied by communications that, by acting on themselves
(= being a function of themselves), produce new communications in the same
domain that can in turn act on themselves and further expand the domain.

It will be clear to the reader that such far-going occupation with self-reference
must change our view of Spencer-Brown's "form" and Niklas Luhmann's sys-
tem/environment and form/medium. Their function lies most certainly not with
the description of the "objects" in their respective domains and their respective
"qualities" as qualities that are eternally true (that is, observer-independent).
Rather, the self-referential construction of the universe and especially the medium of meaning demands the construction of theoretical notions capable of
reflecting themselves as an object (= communication) in their domains, expanding the domain's horizon beyond their own capability of observing that
expansion. Seen in the terminology of topology, form and form/medium are
self-locators or fixed points; they are the sole "points" on the map of mathematics and social theory that coincide with the corresponding point in the terrain
their disciplines are trying to map. Such points contain their own explanation
(that is, their allo-reference and self-reference coincide). They are the pinnacle
of self-reference in domains that are self-referentially built. Therefore, both
LoF and the theory of social systems are not only in the metaphorical sense
a formulation of Quine's paradox. When applied to themselves, they "yield a
falsehood" (absolute because contingent). Yet, therefore, just as Quine's paradox, they can be absolute theories, which are also theories of themselves. Ironically, the latter also constitutes their absolute weakness, I feel. When Niklas
Luhmann, for instance, describes the epistemological premises of his gigantic
theory of society as an invitation to rethink existing social theory and to for-
mulate theories that have themselves compared to his project, this must be
seen as a (rhetorical?) illustration of his epistemological self-confidence, no
more and no less. After all, it is in the nature of meta-theories not to toler-
ate epistemologies of a different brand. Exactly their meta-nature blocks the
possibility of going beyond them—self-reference is infinity in finite guise, as
Kauffman also knows.[69] It should be clear that different theories and different
epistemologies will have to put to themselves the requirement of contingency
and autology in order to qualify as candidates for comparison after all. Whether
this paradox has detrimental consequences is a question that must be left open

here. It is only to hope that, as William Blake has said, "Reason, or the ratio of all we have already known, is not the same that it shall be when we know more."

Notes

Spencer-Brown–related Internet sources:

Laws of Form Web page maintained by Dick Shoup: http://www.lawsofform.org.

Laws of Form Web page maintained by Thomas Wolf: http://www.laws-of-form.net.

スペンサー=ブラウンなんていらない (We don't need Spencer-Brown): Japanese page maintained by Sakai Taito 酒井泰斗; includes links to critical remarks by Kuroki Gen 黒木玄: http://thought.ne.jp/luhmann/sb/Osb.html.

1. I will refer hereafter to the whole of Spencer-Brown's *Laws of Form* as *LoF*; all specific citations, noted as *LoF* with corresponding page numbers, are taken from Spencer-Brown, *Laws of Form*, 1969 (rp. 1994).
2. Von Foerster, "*Laws of Form*." Partial reprints of this review can be found on the back page of several editions of *Laws of Form*; a German translation was included in Baecker, *Kalkül der Form*.
3. Clam, "System's Sole Constituent," 69.
4. See Wagner, "The End of Luhmann's Social Systems Theory"; Zolo, "Function, Meaning, Complexity," and "The Epistemological Status."
5. *LoF*, xi. On mathematics and space, Spencer-Brown has said: "Mathematics is, in fact, about space and relationships. A number comes into mathematics only as a measure of space and/or relationships. And the earliest mathematics is not about number. The most fundamental relationships in mathematics, the most fundamental laws of mathematics, are not numerical. Boolean mathematics is prior to numerical mathematics. Numerical mathematics can be constructed out of Boolean mathematics as a special discipline. Boolean mathematics is more important, using the word in its original sense: what is important is what is imported. The most important is, therefore, the inner, what is most inside. Because that is imported farther. Boolean mathematics is more important than numerical mathematics simply in the technical sense of the word 'important.' It is inner, prior to, numerical mathematics—it is deeper." Spencer-Brown at the American University Masters (AUM) conference, 1973; http://www.lawsofform.org/aum/session1.html.
6. *LoF*, xxviii.
7. Clam, "System's Sole Constituent," 68–69.
8. *LoF*, xi.
9. Varga von Kibéd and Matzka, "Motive und Grundgedanke," 58. A very good illustration of the latter is a series of articles published in a special double issue of *Cybernetics and Human Knowing* (2001), subtitled *Peirce and Spencer-Brown—History and Synergies in Cybersemiotics*. Particularly important when it comes to the direct

relationship between Peirce's notation and the *Laws of Form* is Engstrom, "Precursors to *Laws of Form*."

10. Clam, "System's Sole Constituent," 68–69; compare *LoF*, 96, and Kauffmann: "Self-Reference and Recursive Forms," "The Mathematics of Charles Sanders Peirce," and "On the Cybernetics of Fixed Points."

11. Kuroki constructed an extensive Web site arguing against Spencer-Brown (in Japanese): http://www.math.tohoku.ac.jp/%/7Ekuroki/SB/index.html#abstract.

12. As one sees upon closer examination, both the law of calling and the law of crossing refer to the "closure" or "continence" of the form. On the meaning of continence and the importance of its epistemology, see Varga von Kibéd and Matzka, "Motive und Grundgedanke," 60. Beyond doubt, it contains hints to the form of self-reference, explained in chapters 11 and 12. "A mark or sign intended as an indicator is self-referential" (Kauffmann, "Self-Reference and Recursive Forms," 58).

13. *LoF*, xxix.

14. Ibid., viii.

15. Wittgenstein, *Logisch-philosophische Abhandlung*, § 3.333. In the *Tractatus*'s original:

> Eine Funktion kann darum nicht ihr eigenes Argument sein, weil das Funktionszeichen bereits das Urbild seines Arguments enthält und es sich nicht selbst enthälten kann. Nehmen wir nämlich an, die Funktion F(fx) könnte ihr eigenes Argument sein, dann gäbe es also einen Satz: <<F(F(fx))>> und in diesem müssen die aüßere Funktion F und die innere Funktion F verschiedene Bedeutungen haben, denn die innere hat die Form φ(fx), die äußere die Form ψ(φ(fx)). Gemeinsam ist den beiden Funktionen nur der Buchstabe <<F>>, der aber allein nichts bezeichnet. Dies wird sofort klar, wenn wir statt <<F(F(u))>> schreiben <<(∃φ): F(φu). φu = Fu>>. (Wittgenstein, *Logisch-philosophische Abhandlung*, 34)

16. *LoF*, 1–3.

17. References to Peirce's mathematics are not accidental here: even in notation, the resemblance of Spencer-Brown's form with Peirce's notation (for example, the "sign of illation") is striking. Excellent references here are Engstrom, "Precursors to *Laws of Form*," and Kauffmann, "The Mathematics of Charles Sanders Peirce."

18. *LoF*, 5.

19. Spencer-Brown uses this obscure passage from the first chapter of the *Tao-te-king* as the opening quote for *LoF*.

20. *LoF*, 1.

21. *LoF*, 3. Obviously one must not necessarily agree with this conception, but if so, one places oneself outside the calculus (and we are clearly not willing to do so here). In the calculus, one is expected to obey the injunctions of mathematical communication and *to obey only them and them only*: "In general, *what is not allowed is forbidden*" (Ibid.; emphasis in original).

22. On how *LoF* may aid our understanding of Boolean algebra, see Meguire, "Discovery Boundary Algebra."

23. *LoF*, xxi–xxiii, 87ff.

24. Http://www.lawsofform.org/aum/session2.html.

25. Ibid.

26. Esposito, "Ein zweiwertiger nicht-selbstständiger Kalkül," 99–100.

27. Ibid., 104–5.

28. *LoF*, 68. Therefore, the reference to infinite forms emerges as equivalent to the famous phrase of §7 in the *Tractatus*: "What we cannot speak about we must pass over in silence" (Wittgenstein, *Tractatus logico-philosophicus*, 151). It is the part "in" the calculus that embodies reference to the "outside" of the calculus. Consequently, it has an equivalent problematic status: it is the paradox *saying* that there are *things* about which *nothing* can be *said*. Consider, in this regard, Bertrand Russell's remarks on §70 of the *Tractatus*: "What causes hesitation is the fact that, after all, Mr. Wittgenstein manages to say a good deal about what cannot be said, thus suggesting to the skeptical reader that possibly there may be some loophole through the hierarchy of languages, or by some other exit" (preface to Wittgenstein, *Tractatus logico-philosophicus*, xxi).

29. Luhmann, *Social Systems*, 31.

30. With attention to the linguistic difficulties of describing the unity and difference of distinction and indication, see Schiltz and Verschraegen, "Spencer-Brown, Luhmann and Autology," 65–70.

31. *LoF*, 85.

32. "The fact that men have for centuries used a plane surface for writing means that, at this point in the text, both author and reader are ready to be conned into the assumption of a plane writing surface without question. But, like any other assumption, it is not unquestionable, and the fact that we can question it here means that we can question it elsewhere. In fact we have found a common but hitherto unspoken assumption underlying what is written in mathematics, notably a plane surface. . . . Moreover, it is now evident that if a different surface is used, what is written on it, although identical in marking, may be not identical in meaning" (*LoF*, 86).

33. *LoF*, 99. For examples of Spencer-Brown's engineering work, see http://www.lawsofform.org/patents/index.html.

34. Junge, "Medien als Selbstreferenzunterbrecher," 127. This is not a particularly "exotic" venture: "Die Bestimmung von Räumen aufgrund der Relationen zwischen bestimmten Elementen ist nichts ungewöhnliches: Man konstruiert zum Beispiel eine Landkarte auf Grundlage der Entfernungen zwischen den einzelnen Orten. Diese Entfernungen lassen sich aber auf verschiedene Weisen messen: die Entfernung auf der Luftlinie in Kilometern ergibt ander Distanzen als die Kilometerzahl, die man mit dem Auto zu absolvieren hätte. . . . *Wenn man den Raum operational definiert, wird die am ihm geschulte Logik unter Umständen flexibler*" (127–28; my emphasis).

35. From the discussion below, the reader will understand why "turning up" is the accurate term here.

36. *LoF*, 59ff., 100ff. "To corrupt," synonymous to "to destroy the integrity of" (from Latin *in-tangere*, "untouched," undivided).

37. This topological solution seems to refer to a famous dictum: "The medium is the message" (Marshall MacLuhan). And indeed, there exist some attempts to interpret *LoF* in a medium-theoretical way, such as Lehmann, "Das Medium der Form."

38. The reader must realize that time has thus been *created* as a consequence of a type of space—namely, space in which form can relate to itself and, as such, change (change being the measure of time). Time is thus nothing pregiven. Neither is space: "The first state, or space, is measured by a distinction between states. There is no state for a distinction to be made in. If a distinction could be made, then it would create a space. That is why it appears in a distinct world that there is space. Space is only an appearance. It is what would be if there could be a distinction" (http://www .lawsofform.org/aum/session1.html).

39. *LoF*, 61.

40. Ibid., 65.

41. Schiltz and Verschraegen, "Spencer-Brown, Luhmann and Autology," esp. 65ff. Kay Junge explains this as the medium (of the torus) "blocking" direct self-reference ("Medien als Selbstreferenzunterbrecher").

42. Wagner, "The End of Luhmann's Social Systems Theory," 391.

43. Ibid., 397, 399ff.

44. Luhmann, *Social Systems*, 178.

45. This is not to say that there is a complete and perfect equivalence between the system and the environment; rather, there is a fundamental asymmetry, to the advantage of the system. As a result of this basic asymmetry, the environment does not contain distinctions; it is not a piece of information. Distinctions are found only in a system. Systems observe, while there is nothing self-referential or systemic to the environment.

46. Kauffman, "The Mathematics of Charles Sanders Peirce," 137.

47. *LoF*, 1. As Varga von Kibéd and Matzka stress, continence should be interpreted on the basis of its etymological relationship to the Latin "continere," "to hold together." Thus, "distinction is perfect continence" makes clear not only that the origin of the two sides is to be contained in the distinction, but also (and primarily) that the form has no anchoring in any outside reality or foundation; it is the context of itself, it is "closure."

48. It is important to stress similarities between *LoF* and lambda-calculus. A lambda domain is a class of objects that can act on one another to form new objects of the same kind. It thus presupposes self-reference or recursivity. In "The Mathematics of Charles Sanders Peirce," Kauffman provides most obvious clues on the recursivity of the universe and mathematics in a comparison of Spencer-Brown's ideas with the ideas of Charles Sanders Peirce (102ff.).

49. I surmise this is the meaning and function of the mysterious "unwritten cross," already introduced in the calculus's very beginning: "Suppose any s^n [primal space] to be surrounded by an unwritten cross" (*LoF*, 7).

50. Ibid., 105.

51. Quine, "On What There Is."

52. Clam, "Die Grundparadoxie des Rechts und ihre Ausfaltung," 133.

53. Ibid., 135.

54. Luhmann, "The Autopoiesis of Social Systems," 174; see as well Luhmann, *Social Systems*, 377ff.

55. Luhmann, *Die Gesellschaft der Gesellschaft*, 66.

56. Compare Clam, "Die Grundparadoxie des Rechts und ihre Ausfaltung," 136.

57. Luhmann, "Funktion und Kausalität."

58. Luhmann, *Observations on Modernity*, 44–62.

59. Luhmann, *Social Systems*, 52.

60. *LoF*, 68.

61. Ibid., 106.

62. For a comparison with traditional epistemologies, see Christis, "Luhmann's Theory of Knowledge."

63. *LoF*, 68.

64. Ibid., 101.

65. Ibid. This is, I am afraid, one of the main reasons why *LoF* does have a poor reception. Among others, Kuroki Gen perceives the assumedly mystical nature of *LoF* (especially its numerous forewords) and its cult status with the psychedelic generation of the 1960s as a reason to ignore its insights. Beyond doubt, Spencer-Brown is to a large extent responsible for further isolating *LoF* from a broader academic public, and that is deplorable.

66. Brauns, *Form und Medium*.

67. This leads to the question why Spencer-Brown does not employ a similar view as form as the difference between itself and the medium in which it is written. Peter Fuchs has argued that *LoF* does not need the notion of "medium" as its field of inquiry is mathematics, and as such without heuristical aspirations. As my above considerations on *LoF* show, I am not so sure this is the case. Spencer-Brown makes mention of the notion of medium in at least two passages (see Schiltz, "Form and Medium"). Furthermore, he obviously employs the notion of medium in order to be able to show the possibility of self-referential forms (the medium of plane space versus the medium of the torus). Yet I believe he does not pronounce it in view of his utmost attention for notational matters: "Returning, briefly, to the idea of existential precursors, we see that if we accept their form as endogenous to the less primitive structure identified, in present-day science, with reality, we cannot escape the inference that what is commonly regarded as real consists, in its very presence, merely of tokens or expressions" (*LoF*, 104). In principle, *LoF* can suffice with its topological notation, as long as we are aware of its intricacies (for example, the distinction between crosses and markers in chapter 11).

68. Compare *LoF*, 102ff.

69. Kauffman, "The Mathematics of Charles Sanders Peirce," 105.

Improvisation: Form and Event

A Spencer-Brownian Calculation

EDGAR LANDGRAF

As they say in France, *l'appétit vient en mangeant*, and from our own
experience we might in parody assert, *l'idée vient en parlant.*

Heinrich von Kleist's essay "On the Gradual Fabrication of Thoughts While
Speaking," from 1805–6, explains *en passant* the beginnings of Europe's modern
political and social order as resulting from an improvised speech.[1] In his decla-
ration of June 23, 1789, which led to the fall of the French monarchy, Mirabeau,
Kleist claims, did not yet know what he would say when he first opened his
mouth. After humane beginnings, he did not have "the faintest prescience of
the bayonet thrust with which he would conclude." Only in the process of his
speech did a "fresh source of stupendous ideas [open] up to him," thoughts that
became more and more concrete until they led Mirabeau to leap to the "pinnacle of audacity," asking for the dissolution of the legislature. Kleist concludes
that "perhaps, after all, it was only the twitch of an upper lip, or the ambiguous
fingering of a wrist frill, that precipitated the overthrow of the old order in
France." After a period of disorientation, the speech takes an unplanned and
unexpected direction, developing (seemingly unannounced to the author) its
own dynamics, increasingly lending substance and necessity to itself, until the
speech finally forms an original and now necessary proposition. Kleist's poetic
account of Mirabeau's speech suggests that something new and authentic (here
the "speech act" that presumably changes no less than the political order of the
West) cannot be intended or otherwise derived from a conscious or premedi-
tated act but rather *emerges* from contingencies and improvisation.

The preference for improvisation over premeditation constitutes an aesthetic
principle both for Kleist's literary writings, where embattled protagonists consistently find themselves surrounded by contingencies that force ad hoc action,[2]
and for his aesthetic writings. In "A Painter's Letter to His Son," to mention

another example, the painter explains that in art, the most sublime effect cannot be achieved through planning and premeditation but "may derive from the lowest and most unlikely cause."[3] For Kleist, beauty, art, and grace, but also happiness, love, and justice—in fact, any authentic event—eludes intentionality: it cannot be planned, foreseen, or expected. In the absence of intentions as a structuring principle, Kleist presents contingent and material elements—the twitch of an upper lip, the ambivalent play with cuffs, the mere fact that a person has to speak, etc. and the ability to (re-)act spontaneously—as necessary for the emergence of something new and authentic. Moreover, only contingencies and improvisation can produce events that are of historical significance or that create a singular, aesthetic experience.[4]

Kleist's aesthetics raise the question of artistic creativity—*how* art is created—and suggest that the creative process eludes, even precludes, in some essential way a conscious involvement by the artist. That art exceeds the conscious reach of the artist is not necessarily a new claim. Since ancient times, artists have described the creative act as something transcending their control, a sentiment still shared by many artists today. At the very least, one perceives, as Hans-Georg Gadamer put it, a "qualitative leap between planning and making, on the one hand, and successful completion on the other."[5] In premodern times, this phenomenon was addressed under the heading of artistic "inspiration" or "enthusiasm." Higher powers, God(s) or the Muses, were thought to author the creative process. In the second half of the eighteenth century, those "higher powers" are increasingly internalized. They become part of the creative genius's special abilities. Rather than being perceived as an empty vessel, the artist-genius of the late eighteenth century becomes the model for the exceptional subject, a subject that is able to mediate between the conscious and the subconscious (drives, feelings, nature) or supraconscious (epiphanies, divine insights, inspiration) spheres. Kleist's aesthetics take aim at these late-eighteenth-century figurations of the creative act, seemingly challenging the "cult" established around the artist-genius as an exceptional subject. Insisting on contingencies, accidents, and improvisation, Kleist challenges in particular the anthropocentric viewpoint of these explanations of creativity. Kleist's works highlight a fundamental heterogeneity between artist and artwork, suggesting that the artist merely participates in the creation process through the practice of improvisation rather than authoring the artwork or expressing him- or herself in it in any predetermined manner.

These brief notes on Kleist's aesthetics will serve as a reference and point of departure for the following neocybernetic considerations on improvisation. Drawing on contemporary systems theory and on George Spencer-Brown's form calculus, I will first reflect on the sociohistorical context leading to re-definitions of artistic creativity that encourage improvisation in modern art; second, I will explore the performance aspect of improvisation and its potential effect on the observer. Regarding the first point, neocybernetic discourse suggests alternative descriptions of the art-creating process that can account for the emphasis put on contingency we find in Kleist and in much of twentieth-century art. This implies a basic change in perspective, that we look at the artwork's creation not primarily from the vantage of the artistic individual but as something structured by the particular codes of the art system, codes that secure the reproduction and social autonomy of the system of art.

Since the late eighteenth century, the incorporation of contingent elements and decisions has become a structurally necessary part of the creative process. Kleist's fascination with accidents, coincidences, chance, and other twists of fate reflects this modern involvement of art with contingency, an involvement that will all but define art in the twentieth century. This means in turn that the experiments of Dada, futurism, or expressionism, and later the spontaneity celebrated in beat literature, music, action painting, theater, or performance art, must be viewed as a long-term consequence of aesthetic codes that became dominant in the late eighteenth century. The twentieth-century acceptance of improvisation and contingency in art also signals a new focus on the performance aspect of art, on the production of effects over the communication of meaning.[6] In the second part of this essay I will draw specifically on Spencer-Brown's concept of form and explore its potential to account for the effects of improvisation as performance. This requires a cognitive approach, moving from questions of system differentiation to the question of system irritation. Performed improvisation, as performance art in general, highlights the mutual "irritability" of psychic and social systems. As a functional equivalent of language, art, coordinating cognition and communication, is coupled to the psychic system and can lead the psychic system to reproduce its elements (thoughts, feelings, observations, etc.) in a particular, intensified manner.[7] The question I want to address is how such intensifications are created. Why is art, and in particular why are performance art and performed improvisations, experienced as "events"? How do artworks lead the psychic system to a processing of forms that it experiences as an "event"?

Creativity and Contingency: Historical Considerations

In his insistence on the contingent and material aspects of the creative process, Kleist reacts to the paradoxical situation that artistic creativity faces at the end of the eighteenth and early nineteenth centuries. During that period, art begins to reject set rules and regulations and instead subscribes to codes such as newness, originality, and authenticity to distinguish itself from non-art. These new codes can be understood in the context of the functional differentiation of society, as promoting and sustaining the art system's newly gained autonomy. Artistic autonomy does not mean that art can no longer express social, political, or other interests, but rather that art will define itself and its function(s) according to its own standards without relying on nonartistic factors such as morality, economic feasibility, philosophical truth, the wishes of a patron, etc. to declare what art is and what it is not. The insistence on autonomy leads to arguably the most important development that accompanies the differentiation of the modern system of art, the rejection of the Aristotelian mimesis postulate. Art can no longer be defined in terms of its mimetic function or its representational content, for any representation, strictly stated, will be dependent on what it wants to represent and hence will not be autonomous. That does not preclude representation per se (after all, it will still be a long time before the nonrepresentational arts fully develop) but merely precludes that representational content is used to distinguish art from non-art. The rejection of the Aristotelian mimesis postulate has important consequences for the understanding of the artist and his role in the art production process. In a strict sense, for the creating artist the artwork must be new and original too. It can no longer be a representation of something pregiven to the artist, not even, for example, something conjured up by the artist's imagination, as we would say today. The codes of the modern art system require that the artwork not be planned, foreseen, or otherwise imagined. But how, then, is the artist to conceptualize and create the artwork?

Historically the paradoxical situation in which the artist found him- or herself at the end of the eighteenth century promoted the celebration of the artist as genius and exceptional subject—and this despite (or better *because*) the clear awareness of the paradoxes created for an anthropocentric viewpoint claiming that art cannot be intended. Kant, for example, in his *Critique of Judgment*—one of the first texts that explicitly addresses the question of art from the perspective of its autonomy—defines the artist-genius along a set of paradoxes. According to Kant, the artist must make the artificial look natural, produce something purposeful that nevertheless appears to be without purpose, and act intentionally without seeming intentional.[8] Such paradoxes do not lead the aesthetic discourse to rethink the codes that invite them or to rethink the creative process itself (as I want to do below, drawing on Luhmann's concept of self-programming); rather they lead to an idolization of the artist-genius unheard of in previous times. From a sociological perspective, we could say that like god for religion, the creative genius functions as a "centralized paradox" for the system of art: he is the point where all the apparent contradictions produced by the new codes of the art system are explained in terms of the special abilities that define the artist-genius. Turning paradox into tautology, the artist-genius is who he is because he can do what he can do. He has the ability to intentionally act unintentionally, to lawfully break laws, to allow the artificial to be natural, and to consciously act without awareness. Or as Karl Philipp Moritz, the German writer, psychologist, and aesthetic thinker who articulated the autonomy of art a few years before Kant, put it, the creative genius does not know what he or she is doing yet *feels* that he or she has to do it. The artist sees him- or herself determined in his/her determination of the artwork.[9]

As indicated above, these paradoxes can be understood as responses to the autonomy gained by the art system. The artist "feels" determined in his determination of the artwork because he recognizes that the creation of the artwork follows its own logic. Let us now bring in the sociologist Niklas Luhmann more explicitly and look at his figurations of the art-creating process. In his 1990 essay "World Art," Luhmann distinguishes "object art," which aims at representing existing or imagined objects, from "world art," which no longer represents perceived or imagined worlds but rather calculates (that is, constructs) its own worlds. While object art is in principle mimetic—Luhmann relates it to the cosmological worldview of premodern societies, which he distinguishes from the transcendental worldview that underlies modern society—world art *posits*. It works with the restrictions and options that it derives from its own operations.[10] Luhmann draws on Spencer-Brown's form calculus to describe the creative process in the abstract terminology of distinction theory.

The appeal of Spencer-Brown's *Laws of Form* is its ability to conceive of art in pre-representational terms. In the strict sense employed by Spencer-Brown, distinctions are not signs that denote a preexisting reality, world, idea, or thing, but rather the drawing of distinctions is seen as an operation used by (observing) systems to construct worlds, objects, ideas, things, and also signs.[11] The distinction between signs and distinctions aligns neatly with Luhmann's distinction between object art and world art. While object art relies on signs and symbols, world art constructs its reality through the drawing of and operating with (confirming, condensing, canceling, reentering, etc.) distinctions. After the initial, arbitrary positing of a distinction, the artwork will increasingly be defined by the restrictions and possibilities introduced by each subsequent distinction. The selection of distinctions will in turn be guided by artistic codes.[12] Drawing on Spencer-Brown, Luhmann is able to specify the mechanics and dynamics of artistic creativity as a self-ordering process and detail how order is gained from chaos and necessity from contingency.[13]

In *Art as a Social System*, Luhmann builds on this model and explains the governing of this process and the transition from the contingent to the necessary that it entails by distinguishing between code and program. In a nutshell, Luhmann argues that the functional differentiation of Western society leads to the "self-organization" of modern art. Self-organization means that an operationally closed system uses its own operations to build structures. Art always used *codes* (most importantly the codes beautiful/ugly and interesting/uninteresting) and *programs*. Programs are selection criteria that govern what ensuing operations will be chosen during the production or the observation of art by deciding whether a particular selection belongs to the positive or the negative value of the code. In the eighteenth century, Luhmann argues, code and program separated and began to reorganize the self-organization of art. Especially the demand for newness made it impossible to base one's selection criteria on experience, rules, or existing art; now the artwork had to offer its own program. Luhmann calls the subsequent reorganization of the creative process the "self-programming" of modern art. Self-programming means that the artwork, in the process of its creation, has to develop the program that governs its construction. By developing a program, contingent decisions that mark the beginning of the creative process are increasingly narrowed until the creative process settles into a necessary whole: "Creating a work of art—according to one's capabilities and one's imagination—generates the freedom to make decisions on the basis of which one can continue one's work. The freedoms and necessities one encounters are entirely the products of art itself. . . . The 'necessity' of certain consequences one experiences in one's work or in the encounter with an artwork is not imposed by laws but results from the fact that one began, and how."[14] Only while working on the artwork is the space for decisions defined. All subsequent choices, those that seem necessary as well as those that seem free, are determined by the emerging artwork itself. The concrete, step-by-step realization of the artwork is completed and the artwork a "success and novelty" when the "program saturates, as it were, the individual work, tolerating no further productions of the same kind."[15]

Above I argued that the demand for newness, authenticity, and originality precludes that the artwork can be (fully) planned or foreseen by the artist. Simply put, if the artist is no longer allowed merely to represent what is already there or to follow particular rules or to copy previous models, the creative work cannot follow a predefined plan.[16] The "plan" (program) must emerge during (rather than prior to) the creation of the artwork. Subsequently, each artwork must be constructed from contingent beginnings, requiring that the artist "improvise" (understood in the limited sense suggested by Kleist's short text) until the program for the particular work of art emerges. Drawing on the conceptual vocabulary of systems theory, then, we can understand the need for spontaneity and improvisation as deriving from the operational closure of the system of art. To guarantee newness, authenticity, and originality, and hence to guarantee art's autonomy, the artwork must direct its own completion. Luhmann's model also allows us to understand what Gadamer called the "qualitative leap between planning and making, on the one hand, and successful completion on the other"—namely, as a consequence of the self-programming that each artwork individually must achieve and that cannot be anticipated before its actual completion.[17]

Contingency and Literature

It is important to note that from the neocybernetic point of view, there is no need to hypostasize contingency. Contingencies exist only in regard to particular expectations of necessity—that is, when an observer identifies events that escape or disappoint such expectations (for example, meeting an old acquaintance after many years while crossing the street in the middle of a foreign city constitutes a coincidental event not because each person did not have sufficient cause to be at the particular place at that point in time but because such an encounter was not to be expected). Hence, what is necessary for the artwork need not be for the artist, and vice versa: the artist might well perceive his or her choices as necessary, from the point of view of the emerging artwork; however, they will appear as contingent until they are successfully integrated (become necessary) through the establishing of a particular program. The use, for example, of the color blue might seem important and necessary during a particular period of an artist's life; from the point of view of the painting, however, this initial preference of one particular color must become part of the artwork's program for it to succeed or at least for the color to matter aesthetically.[18]

Insisting thus on a systemic heterogeneity between artist and artworks runs counter to much of the popular understanding of art and literature with its biographico-psychological emphasis on the artist's life, experience, feelings, etc. From the systems-theoretical perspective, art is not the "expression" of the artist at all but can account for the idiosyncrasies of the artist at best as necessary contingencies. What seems "necessary" to the artist from the personal, biographical, political, economic, or even subconscious psychological point of view is, with regard to the artwork itself, mere contingency. In the end, it is the artwork, not the artist, that decides what is necessary and meaningful. Of course, this does not mean that technical expertise and skills, as well as the artist's particular knowledge, would not define important parameters for the emerging artwork; nor does it mean that we will not continue to take interest in the artist and his or her conscious or presumably subconscious motivations. Knowledge about the artist will help us acquire, focus, and apply background knowledge, which may be needed to appreciate the complexity, originality, or oneness of a particular work of art intellectually or emotionally, and thus such information can be important in many regards; it will, however, not define what is artistic about a particular work nor allow us to distinguish art from non-art.[19]

So far in my argument, I have neglected to distinguish clearly between the presentation of contingency *in* art and the presentation of contingency *as* art. Kleist presents contingency on the narrative, semantic, and semiotic level;[20] his writing itself, however, does not follow, for example, the compositional strategies of "spontaneous prose," as Jack Kerouac envisioned and practiced them some one hundred and fifty years later. The majority of modern literature does not follow Kerouac but rather explores contingency on the narrative and semantic level. At least since the early twentieth century (emblematically with Kafka), literature has excelled in creating a sense of contingency for its narratives and for the narrator's and/or character's lives, actions, viewpoints, etc. That is, as modern novels center around characters who are "individuals" because they fail to identify with the norms, views, beliefs, actions, and expectations of their social surroundings (again, think of Kafka or, paradigmatically, of Musil's *Man without Characteristics* or, more radically, of Samuel Beckett's writings), they expose as contingent the observations and observational directives of their social surroundings.

In this regard, contingency distinguishes modern literature from other representations (that is, constructions) of the "world." While scientific, religious, or philosophical representations of the world, for example, must rely on notions of causality, teleology, transcendence, or transcendentality to give their world(s) and worldviews necessity, modern literature seems to thrive on refusing to insinuate "necessity" into its constructions of the world. Its reliance on contingency, however, does not imply a lack of social engagement or relevance. To mention a more recent writer, if we think of W. G. Sebald's artistic interweaving of historical, semi-biographical, and fictional narratives around contingent encounters, events, and photographs, we will find a world saturated with a sense of contingency, yet no less relevant, engaged and engaging in its commentary on Germany's past and present engagement (or lack thereof) with its past.

Luhmann's concept of self-programming suggests that art transforms contingency into necessity. That is, it suggests that the form of modern art is the unity of this distinction, a unity that Luhmann unfolds by temporalizing it.

The unity of this distinction—that is, the reentry of this form within art—also distinguishes modern art from premodern art because premodern Western society going back to Aristotle defined art in contradistinction to reality, as what is possible but not necessary (while reality is necessary). This observational directive ("Look at one as necessary and at the other as contingent") is reentered by modern art's demand that art (the contingent) make the contingent necessary. From a formal point of view, this reentry can take two different forms. On the one hand, art can present *contingent necessity*—that is, present contingent and necessary worlds as necessary. David Roberts's understanding of the form of the novel as the reentry of the distinction between fiction and reality can be read along these lines. Roberts argues that the modern novel as it emerges at the end of the eighteenth century acquires its sense of realism by distinguishing within itself between reality and fiction.[21] On the other hand, art can present *necessary contingency*—that is, contingent *and* necessary worlds as contingent. The latter reentry marks, I would argue, forms of the "fantastic" and the uncanny in literature (and art)—that is, genres or at least moments where the line between what is "real" and what is imagined is blurred and the distinction between the real and the unreal is canceled. In both cases, with the realism of the modern novel and with its fantastic counterpart, the reentry undermines the mono-contextual worldview of the premodern observational directive that insists on the distinction between what is real and what is fiction.

Improvisation and the Modern Differentiation of the Art System

Thus far my historical considerations have used "improvisation" as a heuristic device to explain the creation of autonomous art without having to resort to paradoxical figurations of intentionality such as the ideas of "creative genius" or of a "subconscious." I have neglected the performance aspect of improvisations; today when we speak of artistic improvisation, we assume that it happens on a stage or otherwise before a live audience. Improvisation, defined in this more narrow sense as the *simultaneous conception and presentation of art*, has its own history in music, theater, and especially poetry.[22] In poetry in particular, as Angela Esterhammer has shown, the practice of improvisation as the public performance of poetic creativity was imported from Italy to northern Europe in the late eighteenth century and served both "as a model and foil for an emerging Romantic aesthetics of genius, originality and inspiration."[23] Esterhammer elaborates in considerable detail how Italian performance art both troubled and inspired early-nineteenth-century aesthetics, becoming "a widespread trope for the problematics of spontaneity, performance and

identity as they played themselves out on the international Romantic stage"
until, by the mid-nineteenth century, improvisational performances had acquired a "variety-show aura."[24]

The fortunes of improvisation in art change again in the twentieth century.
Now improvisation is no longer identified with its particular Italian practice
but rather becomes part of art in general—that is, it is practiced in as varied arts
as painting, music, poetry, and theater and (important for such movements)
artistic schools and styles as Dadaism, futurism, expressionism, jazz, Beat literature, or performance art. This development—as well as the recognition of
experimental and improvisational art forms as legitimate contributions to the
art system—signals an increased openness toward contingency. In this regard,
Luhmann's idea of a complete saturation between program and artwork neglects
the role that contingency plays in modern art, where no longer all elements,
aspects, decisions, etc. can be subsumed under a whole. Luhmann's account
seems to be indebted to ideals of "completeness" and "perfection" as we find
them in the aesthetic writings of Moritz or Kant, ideals hardly shared by any
of the more experimental art forms of the twentieth and twenty-first centuries.
This is most apparent with performed improvisations. Here contingency is no
longer confined to the beginning of the creative process but is stimulating and
incorporated throughout the performance. Improvisational theater productions, for example, will rely on continued feedback from the audience and
jazz improvisations on spur-of-the-moment decisions by the performer or the
input of other band members for the continuation of the performance. This
does not preclude the artwork's self-programming per se. In fact, if one did
not expect a program, one could no longer distinguish between successful and
unsuccessful improvisations—that is, one would break with one of the primary
codes of the art system.[25] However, the increased openness toward contingency
challenges the idea of a *complete* saturation between program and artwork (and
subsequently perhaps the idea of the work of art as a closed and complete unit).
Unplanned movements, splashes of color, broken glass, feedback from the
audience, involuntary memories, computer-generated randomness, or maybe
once again the twitch of an upper lip may all flow into the creative process,
continually irritating the self-programming of the artwork.

In opening the art system further toward contingencies, improvisations make
an important contribution to the differentiation of the system of art. From a
systems-theoretical perspective, we could say that improvisation and the overall
incorporation of contingent elements inject "noise" into the system: they provide a resource pool from which the system can draw new impulses to create
new forms and build new structures. Such "noise" does not threaten but rather

enhances the reproduction of complex systems, stabilizing them by making them more adaptable to changes in their environment through the increased variety they can provide. Were a certain art form's ability to be interesting to fade, the art system will already have developed alternatives to take its place. Systems theory, in other words, despite its sometimes conservative reputation, would have us pin our hopes for the stability and continuation of art and its relevance on constant change and increase of variation, rather than promote a return to "proven" artistic models and genres.[26]

Improvisations contribute not only variety to the art system by providing it with a constant flow of new forms. Inasmuch as they are particularly bound to the demands for newness and originality—that is, inasmuch as they are expected demonstrably *not* to follow laws and preexisting patterns—they will also make such laws, patterns, and structures observable to a higher degree. At this point of the argument, we once again need to assume a second-order perspective and observe the unity of the distinction between contingent and planned actions. In improvisation, these are supplemental terms—that is, each side of the distinction defines and preconditions the other. After all, one can improvise—that is, successfully act in unplanned, unforeseen, and unprepared ways—only when one knows what one *cannot* do because it is already known or expected or foreseeable. Such expertise is also expected from the audience. It is one's experience and hence one's expectations for a particular art form and for art in general that will determine what one is able to identify and appreciate as art.

In this regard, the development of the artists/audience, on the one hand, and the artwork, on the other, is structurally coupled. As art changes, the expectations of the artists/audience change, and vice versa; as the expectations of the artists/audience change, art must change. Accordingly, breaking with laws and expectations will not only lead to higher variety, but will also encourage the system to actualize and vary its own laws and structures. Put differently, because improvisations are asked not to conform to the codifications of the art system, they are able to make these codifications visible. Subsequently, improvisations create not only new forms and structures, but also an eye for that which is already old, common, and therefore should no longer count as having artistic value. As a consequence, performed improvisations can be understood as increasing the reflexivity of the system, its ability to observe and connect to its own operations.[27]

What separates the art system from other social subsystems is the variety of forms and the degree of reflexivity at its disposal and subsequently the overall complexity that it can accommodate in its creations. Put differently, while most social subsystems are bound to explain "things" in one way or another and hence (following Luhmann's famous definition) will have to reduce complexity to produce meaning, art can afford to present forms that do not have to explain (things or themselves). Think, for example, of literary representations of historical events, of family dynamics, of madness, etc.—issues that in other fields of study will receive very different treatments. As a result, literature and art are able to accommodate more naturally a much higher degree of complexity.

Improvisation as a Form-Event

After these brief evolutionary considerations of the effects of improvisation on the system of art, I want to turn to the other phenomenon raised by Kleist's short text on Mirabeau's speech—namely, to the event-character of improvisation. Simply put, I want to address the question of how performed improvisations captivate their audience. In particular, I am interested in the aesthetic experience during the actual performance, rather than in the possible adoration for the artist or the artwork that might follow and feed one's interpretive desires. Raising such questions is in line with the general shift in art toward the production of effects. Although in recent years this shift has finally received wider attention in academic circles, we still lack a critical vocabulary that would allow us to describe the performance character of improvisations rather than merely interpret its potential meaning or describe the particular techniques employed.[28] The absence of such a critical vocabulary is especially regrettable with regard to improvisation because here the event-character is an essential part of the simultaneous conception and presentation of art.

Seen as events, improvisations, in a strict sense, are nonrepresentational, nonmeaning producing acts. That does not imply that improvisations cannot represent something or carry a particular meaning. However, neither what is represented nor its potential meaning distinguishes improvisation from other creative activities or can define the particular effect created by a performed improvisation, the temporal and sensual immediacy of the act. The simultaneous conception and presentation of art appears to be able to captivate an audience and create an "experience of presence" (Gumbrecht), independent of any meaning that subsequently might or might not be attached to such a performance. To some extent, such an "experience of presence" might be created by all performances and perhaps by all art that "succeeds"—that is, art that is able to present itself as an object or act that draws the attention of the observer in a uniquely intensified and intensifying manner.[29] Nevertheless, only in improvisation and performance art does this effect become the primary goal of the creative act. Because of their fleetingness, they must be able to draw the attention to the here and now.

How is this done, and how can it be described? To answer these questions, I want to build on a 1999 essay by Hans Ulrich Gumbrecht entitled "Epiphany of Form: On the Beauty of Team Sports." I am particularly interested in Gumbrecht's use of "form" to describe what he calls the "production of presence" in American football. I should note that the analogy of sports and art is not based on the circumstance that in sports, too, we find improvisation in the sense of unplanned and unexpected actions; rather, I am interested in the fact that sports events share with improvisations the central characteristics of successful (noted and notable) performances—namely, the focus they command on the here and now. The subsequent analysis of a sports event or a performed improvisation might be interesting, even intriguing, in terms of its mesmerizing effect; however, it will never compare to what people experience when they observe an actual artistic performance or a sports match, a live race, their team competing in a soccer tournament, etc.[30]

As Gumbrecht and others have argued (see also, for example, Fischer-Lichte), hermeneutic approaches to such phenomena—that is, approaches that search for meaning or symbolism in such events—as interesting as they might be, miss precisely this effect. Taking what he calls the "blind spot" of hermeneutics (that it always operates in the realm of representation) as his starting point, Gumbrecht locates the "Other of mimesis" in the production of presence. Gumbrecht defines presence not as something given or self-evident in the Husserlian sense of the word and as Derrida liked to deconstruct it, but rather as "form-as-event," as "the convergence of an event-effect with an embodied form."[31] "Event" Gumbrecht defines simply as a singular and contingent action; "form" he defines as twofold—on the one hand (following the musicologist Gerhard Berger), as a "movement whose direction we want to see continued," and on the other (following Luhmann's preferred reading of Spencer-Brown's form concept), as the unity of a distinction. With the latter, we have to understand form as a distinction that constructs what it makes observable in the first place.

By drawing on Spencer-Brown's calculus, Gumbrecht escapes an ontological worldview but also the danger of creating a presence category that could be interpreted in terms of religious immanence (an accusation he refutes at length in his more recent book on the *Production of Presence*). As indicated above, Spencer-Brown's form concept is useful here because it is not representational, but rather operational, letting complexity emerge from a simple and contingent beginning.[32] Because of this pre-representational quality, the form concept lends itself to describing events that cannot be accounted for in terms of their representational properties. This is the case with the event-character of a sports match or performed improvisation: they are not about anything that would preexist the event itself; rather, the event, as Gumbrecht puts it, is about its own emergence, about the original emerging of forms. Gumbrecht, reading with Luhmann a strongly cognitive dimension into the form concept of Spencer-Brown, argues that this originating process captivates the attention of the audience by confronting the audience with forms in the Spencer-Brownian sense—that is, with the unity of distinctions. American football, for example, confronts the audience with a playing field that is first empty and then filled; with players that are once frozen then move; with situations where nothing happens but something is about to happen; where someone is here and then there, etc. For team sports, the distinction between entropy and negentropy, chaos and order, is especially important. While the defense fights for order, the offense attempts to create chaos for the defense and hence has to act in an unforeseeable and surprising fashion, and *vice versa*. The tension, but also the excitement for the audience, is greatest when the offense is able to surprise through unforeseen and unprepared-for actions or when the defense is able to react to such a challenge successfully.

Artistic improvisations, I want to argue, aim in comparable ways to captivate the attention of the audience and to produce presence. Improvisations stage the emergence of new forms as an event: they are singular, momentary, and contingent. Singularity further heightens the tension because what is presented cannot be taken back. Every moment counts; every decision is final and therefore of increased significance.[33] Improvisations also create "forms" according to both of the definitions Gumbrecht uses. Corresponding to Berger's definition of form, improvisations create a movement whose direction we want to see continued. In art, such a desire for continuation can be explained in terms of Luhmann's concept of self-programming: the increased narrowing of possible choices directs the continuation of the creative process as well as the spectator's attention and expectation. Past selections decide what subsequent selections are desirable or not, possible or not. Jazz musicians have long had an eye for this effect. Paul Berliner, in his extensive study of the different techniques used in jazz improvisation, quotes Max Roach describing his experience with improvisation: "After you initiate the solo, one phrase determines what the next is going to be. From the first note that you hear, you are responding to what you've just played: you just said this on your instrument, and now that's a constant. What follows from that? And then the next phrase is a constant. What follows from that? And so on and so forth. And finally, let's wrap it up so that everybody understands that that's what you're doing. It's like language: you're talking, you're speaking, you're responding to yourself. When I play, it's like having a conversation with myself."[34]

Based on similar observations, Ted Gioia suggests that we distinguish two creative methods to separate the creative process in traditional art from improvisations. While traditional art follows what Gioia calls the blueprint method (that is, it plans ahead), improvisations use the "retrospective method." According to the latter, "the artist can start his work with an almost random maneuver—a brush stroke on a canvas, an opening line, a musical motif—and then adapt his later moves to this initial gambit."[35] Such descriptions understand improvisation as a complex feedback process that builds forms out of contingent elements by relating present decisions to past decisions. The improvisation responds to itself, repeating and altering, changing and rephrasing what has come before. Berliner describes in great detail the many possibilities jazz improvisers have in "responding to their own notes"—for example, "by pausing briefly after an initial statement, then repeating it, perhaps with minor changes such as rhythmic rephrasing. This also allows time for the player to conceive options for the subsequent phrase's formulation.... Artists may 'run' the figure directly 'into itself,' perhaps through a slight extension or short connection pattern, treating the figure as a component within a longer phrase.... Experienced improvisers can create variations on each feature of extensive melodic-rhythmic material within the framework of lengthy antiphonal phrases."[36]

Without drawing on the neocybernetic discourse, Berliner describes improvisation as a computation of complex forms. Improvisations draw, condense, confirm, cancel, and compensate distinctions. This computation of forms constitutes an evolutionary process that is a process of variation, selection (program), and stabilization. Performed improvisations thus process and relate entropy and negentropy, chaos and order, structured and unstructured, prepared and unprepared, known and unknown elements. In jazz, for example, improvisations often start with or allude to or culminate in known melodies or harmonies that soon after are decomposed again. The effect of surprise especially hinges on resorting to and then changing the familiar. Theatric improvisations similarly work with known elements, familiar quotes, characters, plots, situations, etc. just to change them around again in each performance.[37]

This play with forms, I would argue, is designed and able to draw our cognitive attention. The observer wants to know what will happen next, which distinctions will be condensed, confirmed, canceled, or compensated. In art, it would seem, such expectations will be created and directed both by the general codes of the art system (for better or worse, we visit concerts or museums with different expectations than we have for the dentist's office) and by the program that the work of art or performance develops for itself. The latter presents a particular challenge that is distinct from the knowledge and experience needed when we watch, for example, a sports event. While the sports connoisseur has to learn the rules and parameters of a sport and know its major figures and internal dramas, the process of "recognition" is more intricate in art. Here the "rules" are not known in advance. The artwork must develop a unique program for itself. Improvisations, and performance art in general, challenge the observer to recognize the unique program as it emerges for the first time. If the observer succeeds in identifying such a program (the simplest example might be that of a particular rhythm), the self-programming of the performance will become more and more compelling and make an identification with the observed process possible.

Identification here would mean that temporarily one would no longer observe the action from a distance—that is, draw one's own distinctions to observe the event—but instead would "embody" the process. That is, to experience a performance as such, one has to experience the logic of the distinctions drawn and the operations performed as one's own. The point is that such an identification with the performance is not symbolically mediated, as Adorno would have it, but rather results from a (nevertheless cognitive) identification with the self-programming of the form-creating process. In jazz, such forms of embodiment are commonly referred to as "feeling it" or as "getting into the groove," as states of affectedness that compel artist and audience to identify with the ongoing performance. The experience is intensified—that is, tension is created and attention drawn—by the particular temporal structure of performances. Confronted with forms (the unity of distinctions), time appears to expand: the observer waits for the tension to be resolved, for the play to start, the performance to continue, the melody to reappear, the painting or sculpture to take shape, the program to settle, etc. The observer's re-tension, where time seems to expand, is opposed to the moment of decision and recognition, which focuses the attention of the observer alone on the here and now. During artistic improvisations, re-tension and decision do not simply follow each other; rather both are simultaneously processed, captivating the observer's attention continually between re-tension and decision.

Materiality and Form

One of Luhmann's most controversial and yet fundamentally important theorems concerns the separation between consciousness and communication as two independently operating closed systems that cannot determine each other's operations. At first glance, this basic theorem seems incompatible with ideas of presence or embodiment as I have tried to describe them above. How can art or a performance—social phenomena—create a particular state of consciousness? Per se, they cannot, of course. As all of us who have fallen asleep at the opera, were bored in a museum, distracted during a concert, or simply annoyed by a particular performance will know, no work of art and no performance can determine if and how it will be perceived by the psychic systems it finds in its environment. Art can only offer structures that the psychic system will or will not (be able to) adopt. Furthermore, the ability to appreciate and experience art or any performance is acquired. A person who has never listened to jazz before will not be able to appreciate the subtleties and innovativeness (or lack thereof) of a particular performance. Nor will he or she be able to embody the self-programming of the emerging work or a particular improvisation. As indicated above, even team sports require a considerable amount of pre-knowledge (rules, characters, and narratives) for the event to be "lived." Art, because of its self-programming, requires, I would argue, a higher degree of readiness and a more refined form of expertise. Yet even when one is fully prepared and familiar with a particular art form, cognitive identification with the process will be momentary and sporadic and can by no means be guaranteed.[38]

I make this point to underline that any "experience of presence" is produced by the internal operations of the psychic system. Art can irritate the psychic system and offer particular observations to it, but it cannot determine or define its operations. Put differently, the psychic system must learn how to react to the irritations of art. However, this does not mean that the appreciation of art and the "experience of presence" can take place only on the conscious or intellectual plane. This is where Spencer-Brown's form concept is most useful. It allows us to describe the observational modes of systems without having to presuppose (or exclude, for that matter) meaning or intellect from the observational process. In this regard, Spencer-Brown's form concept presents an alternative or the possibility to "unpack" recently redefined conceptions of "materiality" or "embodiment" that are based on existentialistic or ontological definitions of the human body and our being-in-the-world and that are often employed to approach an aesthetics of the performative. Erika Fischer-Lichte, for example, finds the term "embodiment" central for her aesthetics of the performative. She defines embodiment, however, in rather existentialistic terms. Following the anthropologist Thomas J. Csordas, she wants to "grant the body a comparably paradigmatic position as is granted to text." The term "embodiment" should "open a new methodological field, where the phenomenal body, the physical being-in-the-world of man, figures as the condition for the possibility of any cultural production."[39]

In an article from 1996, Gumbrecht explicitly rejects the form-without-matter concept proposed by Luhmann's and Dirk Baecker's reading of Spencer-Brown, arguing that "forms which are events, forms which are 'being born to presence' . . . cannot occur without matter or substance—nor be analyzed without concepts of matter and substance."[40] Gumbrecht's most recent book on the subject, *Production of Presence: What Meaning Cannot Convey*, completely omits Spencer-Brown's or Luhmann's conception of form and instead draws (among other writers) on Heidegger and on semiotic considerations (especially of transubstantiation) to explore moments of presence where our "being-in-the-world" is noninterpretive, not primarily based on meaning, and in this regard seems to overcome, albeit only temporarily, the separation between thought and being that defines our Western metaphysical culture. Such attempts to define materiality and presence in art in semi-existential or ontological terms hope to find an alternative to the culture of thought that dominates (academic) Western culture by grounding the authenticity of human experience in the materiality of the human body.[41]

I want to suggest that drawing on systems theory and on Spencer-Brown's pre-representational form concept allows us to overcome the perceived division between materiality and body, on the one hand, and a culture of thought on the other. Neocybernetic discourse allows us to understand the "experience" created by a person's cognitive engagement with art without having to assume an interpretive stance toward the work of art or performance. This means that we redefine cognition beyond the traditional subject/object distinction and without linking it to the production of meaning, replacing the former with the system/environment distinction and the latter with Spencer-Brown's form concept. With the help of such conceptual substitutions, we can comprehend the psychic and the nervous systems as observing and relating to their environment long before comprehension mediated through language and abstraction is initiated.

At crucial points of my argument, I have relied on the concept of "expectation" to understand how the psychic system engages a performance—that is, how it processes irritations from the art system and how performances create an intensified experience. Expectations create tensions, resolutions, surprises, disappointments, etc. and thus structure our attention in particular ways, engaging us not only intellectually but also emotionally. In addition, expectations are a helpful concept in describing experience because they do not have to be conscious and yet, for the most part, they will be acquired. The nervous system, for example, might have been trained to "expect" a certain stimulus (smell, sound, etc.) upon entering a room, but will only react to or notice its own expectations when they are not met. Thus, performances focus attention and create events by addressing and creating (confirming, raising, disappointing, exceeding, etc.) expectations of which one might or might not be aware. The performance or appreciation of a jazz improvisation surely does not require— for the moment of its performance even *prohibits*—the careful analysis of the creative process; yet a high level of "expertise" entailing numerous conscious and unconscious expectations will be at play. Systems theory explains such phenomena—the psychic system's ability to relate to the social system and *vice versa*—in terms of structural coupling. Today the cognitive sciences can explain the creation of unconscious expectations as resulting from the acquisition of specific cultural skills through the formation of neural networks.[42] As Dreyfus and Dreyfus remark, neural networks "provide a model of how the past can affect present perception and action without the brain needing to store specific memories at all."[43] Through understanding and practice, processes are embodied—that is, they can ultimately be performed without conscious or thoughtful involvement.[44]

Two important points follow from these considerations. For one thing, the conceptual models used here help us not to hypostasize notions of "presence," "materiality," or "embodiment" as, I believe, Gumbrecht and many others are in danger of doing when they rely on notions of materiality and substance (rather than distinction and indication) as preconditions for the possibility of experience and observation. Such notions make it difficult to account for the expertise required for both the creation and the appreciation of art. Drawing on neocybernetic discourse, we can instead acknowledge the high degree of prior knowledge and exposure needed for the creation of an "experience of presence" (even in ritualistic cultures, I would argue, "preparation" for the event is needed) without having to assume that this experience is based (solely) on conscious calculations and understanding. This is not to overlook the difference between the experience created by a live performance and the experience of reading a book. The difference, however, is perhaps less one of quality than of quantity. In the performing arts and in improvisations, the play with conscious and unconscious expectations will take place on multiple levels simultaneously. Performances not only create an immediate confrontation with forms and force special attention because of the singularity and irreversibility of the process, but they also "surround" the psychic and nervous systems, irritating them on multiple levels. This creates situations that cannot be replicated by books, museums, or lectures or even by virtual realities where eyes, ears, and maybe even our directional senses are stimulated in a predetermined fashion. The role of the body's presence, then, would be defined and limited by the totality of observations and expectations it can elicit. Systems theory and Spencer-Brown's form concept could help us analyze in more detail (than was possible in this essay) the complex of possible irritations of the nervous, psychic, and social systems and the embodiment and experiences of presence made possible through their interaction.

Notes

1. For the quote in the epigraph and the following quotes from "On the Gradual Fabrication of Thoughts While Speaking," see Kleist, *An Abyss Deep Enough*, 218–20.
2. Think of Penthesilea's famous "misunderstanding," confusing bites and kisses because in German both words happen to sound similar; or the trial by ordeal that drives the story in *The Duel*. Tragedy in Kleist's texts often results from the attempt to attribute meaning (a cause, purpose, volition) to encountered coincidences.
3. Kleist, *An Abyss Deep Enough*, 240. The rejection of premeditation as a creative principle finds its perhaps most radical articulation in Kleist's famous essay "On the Puppet Theater," with its claim that the ends of the world meet, that the utter absence of consciousness is equal to infinite consciousness, equating the marionette with god.
4. Wellbery reads Kleist's writing as elaborating and enacting narrative strategies that maximize the opportunities for the contingent and accidental to occur. Kleist's "On the Gradual Fabrication," Wellbery argues, elaborates the alterity of language, how in the process of speech production "an alteration takes place through which something unforeseen, even by the subject producing the speech, occurs" ("Contingency," 245). The alterity of language, that "the text escapes my intentional control," is, according to Wellbery, a structurally necessary ingredient for an utterance to become an event: "What Kleist has done, then, is to introduce chance into the production of speech itself. Contingencies do not merely come to language from the outside but rather are effective, as it were, from the beginning, within the individual utterance. By virtue of that fact (which is the fact of facticity), the utterance can become an event, the site of the emergence of the new" (ibid.).
5. "Es ist ein Sprung zwischen Planen und Machen einerseits und dem Gelingen" (Gadamer, *Die Aktualität des Schönen*, 44; translation here and in the following mine). "Sprung" means both leap and crack, split. According to Gadamer's highly perceptive analysis—which I am using here and in the following to demarcate some of the parallels and differences between the anthropocentric, hermeneutical approach to art and its neocybernetic reconceptualization—it is this "Sprung" between making and completion that "distinguishes the artwork in its uniqueness and irreplaceability." Gadamer relates this leap to what Walter Benjamin called the aura of the artwork: "Es ist ein Sprung, durch den sich das Kunstwerk in seiner Einzigartigkeit und

Unersetzbarkeit auszeichnet. Es ist das, was Walter Benjamin die Aura des Kunst-werkes genannt hat" (ibid.).

6. Fischer-Lichte in her important study, *Ästhetik des Performativen* (Aesthetics of the performative), uses the term "performative turn" to describe the often noted de-limination (*Entgrenzung*) of art in the 1960s as a common tendency of different art forms to "realize" themselves in their performance: "Die immer wieder beobachtete Entgrenzung der Künste seit den sechziger Jahren des 20. Jahrhunderts läßt sich also als performative Wende beschreiben. Ob bildende Kunst, Musik, Literatur oder Theater—alle tendieren dazu, sich in und als Aufführung zu realisieren" (*Ästhetik des Performativen*, 29).

7. See esp. chapter 1, "Perception and Communication: The Reproduction of Forms," in Luhmann, *Art as a Social System*.

8. "Therefore, even though the purposiveness in a product of fine art is intentional, it must still not seem intentional; i.e., fine art must have the look of nature even though we are conscious of it as art. And a product of art appears like nature if, though we find it to agree quite punctiliously with the rules that have to be followed for the product to become what it is intended to be, it does not do so painstakingly. In other words, the academic form must not show; there must be no hint that the rule was hovering before the artist's eyes and putting fetters on his mental pow-ers" (Kant, *Critique of Judgment* [1987], 174 [§45]). In the subsequent paragraphs of the third *Critique*, Kant describes the special "talents" of the genius who is able to resolve these paradoxes—that is, "through which nature gives the rule to art" (174 [§46]).

9. See "Über die bildende Nachahmung," in Moritz, *Werke*, 2:564ff. Such paradoxes lead the aesthetic and the emerging anthropological discourses to expand their defini-tions of the "unconscious" as an area where unintentional intentions, unlawful laws, unnatural nature, etc. house. For a more detailed analysis of Moritz's artist-genius, the paradoxes it involves, and their unfolding, see Landgraf, "Self-Forming Selves."

10. Luhmann, "Weltkunst," 23.

11. As Varela put it, "By going deeper than truth, to indication and the laws of its form, [Spencer-Brown] has provided an account of the common ground in which both logic and the structure of any universe are cradled, thus providing a foundation for a genuine theory of general systems" ("A Calculus for Self-Reference," 6). The difference between truth claims and the form calculus is that statements about truth, the world, reality, etc. operate with signs that at some point must be understood as representing (or failing to represent) what they attempt to designate, while Spencer-Brown's calculus, operating with distinction and indication, constructs what they observe (on this point see also Simon, "Mathematik und Erkenntnis," esp. 55ff.). For a more extensive discussion of the epistemological and ontological implications of Spencer-Brown's *Laws of Form* and a response to some of its critics, see Schiltz in this volume.

12. Luhmann, "Weltkunst," 18–23.

13. Luhmann's model, of course, is not the only one that mirrors Kleist's description of Mirabeau's speech cited above. Below, I will quote a number of examples from jazz musicians who describe the creative process in very similar terms.

14. Luhmann, *Art as a Social System*, 203–4.

15. Ibid., 202. The idea of a complete saturation of program and artwork mirrors Moritz's formulation that true artworks cannot be described because they are perfect descriptions of themselves ("weil sie die vollkommensten Beschreibungen ihrer selbst sind" [*Werke*, 2:587]). The proximity of Luhmann's model to late-eighteenth-century formalizations of art raises the question if this model can be used to describe the provocations of twentieth-century and contemporary art. I will return to this question below.

16. The point is not that the artist cannot predetermine themes, ideas, forms, media, perspectives, styles, etc. that he or she wants to use or comment on with a work of art; it is merely that such predetermined factors remain elements that the artwork uses for its self-programming. It goes without saying that Luhmann's conception of art is more comprehensive than the one aspect I single out here in order to focus on the role of contingency and improvisation in the creative process for modern (rather than premodern) art.

17. Spencer-Brown understands his calculus in similar terms as self-referentially closed— that is, as a construction that creates rather than represents a world. He notes that the calculus develops its own program, which is valid only for this particular work. In the last paragraph of his introduction to the 1969 edition of *Laws of Form*, he describes his position as author of *Laws of Form* in the artistic terms suggested above. Spencer-Brown rejects almost all authorial responsibilities, claiming that what he wrote down "wrote itself," that after a contingent beginning, he was merely responsible for the instrumental labor of constructing a record that he hopes is clear and concise. In the preface to the 1994 edition of *Laws of Form*, Spencer-Brown relates this experience directly to art. Citing the neo-empiricist Herman von Helmholtz, he makes the point that the artistic process constructs "existence" independent of the intentions of the artist. Despite its self-programming, Spencer-Brown's *Laws of Form* are, of course, not a work of art. While we might consider them to be "interesting," even "beautiful" according to certain artistic standards, they certainly lack the very sense of contingency that characterizes much of modern art.

18. The example raises another question (which is beyond the purview of this essay)— namely, the question of style, the ratio of continuity and variation between different yet related works of art.

19. This conclusion is, of course, not unique to systems theory. For example, despite taking a decidedly anthropocentric viewpoint, Gadamer concludes his essay on the "Relevance of Beauty" by remarking that the "perfect experience of an artwork is such that one finds oneself adoring the total discretion of the actors: that they do not promote themselves, but instead evoke with almost involuntary matter-of-courseness the work, its composition and its internal coherence" (Die vollendeste Erfahrung eines Kunstwerkes ist so, daß man gerade vor der Diskretion der Akteure

mit Bewunderung steht: daß sie sich nicht selbst zeigen, sondern das Werk, seine Komposition und seine innere Kohärenz bis zur ungewollten Selbstverständlichkeit evozieren) (*Die Aktualität des Schönen*, 69).

20. The last has received much attention from deconstructive readings of Kleist's work. Hamacher's article "Das Beben der Darstellung," for example, follows Kleist's de- and re-composition of the word *Zufall* (coincidence) in a semiotic, semantic, and narrative net, where *Fall, fallen, zusammenfallen* (fall, falling, falling together), etc. combine in ways that stage coincidence as a "linguistic event" that "brings to fall" the representational function of speech ("Das Beben der Darstellung," 156–57).

21. "The novel authenticates itself by foregrounding the difference between fiction and reality within the fiction. In other words, in contrast to traditional narratives, the question of realism (external reference) is defined in terms of the unity of the difference between self- and external reference" (Roberts, "Self-Reference in Litera- ture," 42). Clarke shows how such reentry figures shape Shakespeare's *A Midsummer Night's Dream* (for that very reason, we should consider it a "modern" play) and points (with Roberts) to the second-order observations that are made possible by such maneuvers, making literature in particular "an epistemological device for in- terminably deferring the location of an ultimate perspective from which the be- ing of things could be thought to be known once and for all" (Clarke, *Posthuman Metamorphosis*, 71).

22. I am following the definition found in Finscher, *Die Musik in Geschichte und Ge- genwart.* The encyclopedia offers a lengthy entry on the history of improvisation in music.

23. Esterhammer, *Spontaneous Overflows and Revivifying Rays*, 9. It is safe to assume that Kleist was familiar with the first comprehensive study of the nature, history, and aesthetics of improvisation—namely, Fernow's "Über die Improvisatoren" (On the improvisers), first published in 1801 in the *Neue Deutsche Merkur.* The text portraits the Italian art of improvisation in terms of an aesthetics of genius and enthusiasm, emphasizing not only the technical skills needed or the spontaneity of presentation, but in particular the immediacy of this poetry, its ability to "inspire directly" the soul of the listener (see esp. 304).

24. Esterhammer, "The Cosmopolitan Improvvisatore," 157.

25. In *Die Aktualität des Schönen*, Gadamer addresses improvisation only once—namely, to explain how the unity of the artwork is established through a hermeneutic act that constitutes the identity of the work by understanding it as a unit. Improvisations show this most effectively. Despite their fleetingness, we recognize them in their existence as "works"; otherwise we could not judge their quality or distinguish them from mere finger exercises (33).

26. From this perspective, we need not share the cultural conservatism expressed, for example, by Gioia's concern for the future of jazz. Noting an increased fragmenta- tion of styles, Gioia fears that "the benefits of pluralism threaten to collapse into the uncertainties of relativism" (*The Imperfect Art*, 74). For all its important contribu- tions to the study of jazz, Gioia's book clearly lacks concepts that would allow him to

approach the improvisational art of jazz as part of a broader tendency within modern art's evolution. Instead of challenging the aesthetic tradition whose concepts fail to account for the specificities of this improvisational art form, Gioia propagates an understanding of jazz in terms of nineteenth-century aesthetics of genius that asks us to ignore this art form's "imperfections" and appreciate improvisation as "the purest expression possible of the artist's emotions and feelings" (83).

27. If one considers how much improvisation has become part of the modern performing arts and how much the technical and material aspects of art production have been exposed in the twentieth century, it must come as a surprise that the idolization of the artist-genius has not diminished. At least within the popular discourse on art, the idea of the artist-genius remains a viable concept to explain the creative process. Even artists who explicitly celebrate contingency, materiality, or the technical aspect of the creative process seem to profit from the myth of the genius. One need only think of the aura that surrounds Bebop greats such as Charlie Parker or John Coltrane, the recent movies celebrating Jean-Michel Basquiat or Jackson Pollack, the legendary status attributed to Josef Beus, etc. The continued popularity of such improvisation artists cannot be explained simply on the basis of their exceptional skills. Their star status and its association with the idea of genius are more likely a result of the modern-day reception of art by the mass media. The modern media review art almost exclusively from an anthropocentric perspective. It is safe to assume that the art system is not protesting against this misrepresentation because artists and the art system's institutions profit economically from the commercialization of its heroes.

28. Fischer-Lichte argues that into the late 1980s, culture was understood almost exclusively as text. Only in the 1990s does a change emerge toward an understanding of culture as performance that allows the academic pursuit of performance art and of performative aspects within art in general (see esp. *Ästhetik des Performativen*, 36ff.).

29. Citing the example of what to him seemed to be a cloth rack that was offered with great effect as a work of art, Gadamer argues that "the creation of effects is the primary calling of contemporary art" (In seiner Wirkung und als diese Wirkung, die es einmal war, hat es seine Bestimmtheit) (*Die Aktualität des Schönen*, 33).

30. Gumbrecht suggests that one compare this experience with Nietzsche's conception of the Dionysian experience. They seem to be able to breach at least temporarily the separation between subjects and the separation between subject and object world, between observer and observed ("Epiphany of Form," 366).

31. Ibid., 359.

32. See Schiltz in this volume for a more extensive discussion of the connection between operationalism and form. As he points out, Spencer-Brown's operationalism allows us to account for self-referentially operating systems "consisting of operations that take their own results as a base for further operations. These are forms that 'in-form' themselves" (165). The concept of "in-formation" as developed by Schiltz seems especially apt to describe the form finding/creating process that marks the self-programming of art.

33. "The improviser makes a succession of choices in performance which cannot be erased, so everything (s)he does within the performance must be incorporated into the whole. This involves an attentiveness to the present moment, so that creativity is a response to the here and now, though the choices made by the improviser are inevitably influenced by past experience of improvising" (Smith and Dean, *Improvisation*, 26).

34. Berliner, *Thinking in Jazz*, 192.

35. Gioia, *The Imperfect Art*, 60.

36. Berliner, *Thinking in Jazz*, 193–94.

37. Fischer-Lichte sees the creation of a special relationship between actors and audience as the central function of performance art and understands the modern (post-1950s) openness toward contingency as expressive of the explicit interest to create "feedback loops" and thus "self-referential, autopoietic systems that can no longer be interrupted or controlled." That is, she recognizes a "shift from the possible control of a system to a special modus of autopoiesis" in theater productions since the 1960s ("Mit der performativen Wende in den sechziger Jahren ging eine neue Haltung gegenüber der Kontingenz einher. Sie wurde nun überwiegend als Bedingung der Möglichkeit von Aufführungen nicht nur akzeptiert, sondern ausdrücklich begrüßt. Das Interesse richtete sich explizit auf die *feedback*-Schleife als selbstbezügliches, autopoietisches System mit prinzipiell offenem, nicht vorhersagbarem Ausgang, das sich durch Inszenierungsstrategien weder tatsächlich unterbrechen noch gezielt steuern läßt. Dabei verschob sich das Interesse von einer möglichen Kontrolle des Systems zu einem besonderen Modus der Autopoiesis" [*Ästhetik des Performativen*, 61]).

38. Despite his expertise regarding jazz performances, Gioia nevertheless adds the following: "[That] much—if not most—jazz is boring seems scarcely undeniable" (*Imperfect Art*, 109)—a statement that, I would claim, speaks more to the difficulty of psychic systems to sustain the experience of presence without distraction than to the intricacies of jazz as an art form.

39. "Es geht ihm also darum, dem Körper eine vergleichbar paradigmatische Position zu verschaffen wie dem Text, anstatt ihn unter dem Textparadigma zu subsumieren. Das eben soll der Begriff *embodiment*/Verkörperung leisten. Er eröffnet ein neues methodisches Feld, in dem der phänomenale Körper, das leibliche In-der-Welt-Sein des Menschen als Bedingung der Möglichkeit jeglicher kultureller Produktion figuriert" (Fischer-Lichte, *Ästhetik des Performativen*, 153; my translation).

40. Gumbrecht, "Form without Matter," 587.

41. For an excellent historical account of notions of body and embodiment in anthropological theory, see Csordas, "Embodiment and Cultural Phenomenology."

42. Hubert and Stuart Dreyfus analyze Merleau-Ponty's concept of embodiment drawing on the model of neural networks: "According to these models, memories of specific situations are not stored. Rather, the connections between 'neurons' are modified by successful behavior in such a way that the same or similar input will produce the same or similar output" ("The Challenge of Merleau-Ponty's Phenomenology," 115).

Like Gumbrecht, they relate the acquisition of expertise to the experience described by athletes: "When everyday coping is going well, one experiences something like what athletes call flow, or playing out of their heads. One's activity is completely geared into the demands of the situation" (ibid., 111).

43. Ibid., 115.

44. Dreyfus and Dreyfus distinguish various stages of skill acquisition from the novice to the expert. Experts—chess players, drivers, pilots—no longer need to think through complex decision-making processes but anticipate and act intuitively.

Communication versus Communion in Modern Psychic Systems

Maturana, Luhmann, and Cognitive Neurology

LINDA BRIGHAM

Contemporary technological society depends on securing consensus about time. Social complexity demands the ordering of a huge array of actions carried out over a distance too great for direct communication. Timing compensates for the absence of communication; temporal measures independent of individual experiences form structures against which activities can be synchronized without direct connection. The demands of global timing also require the transcendence of merely local or natural temporal markers such as harvest, tides, sun, and moon. Since the sixteenth century, these ancient rhythms have been progressively displaced by abstract, precise, and portable measures to accommodate modern mobility, dispersion, and multiplicity of viewpoints.

The necessity of temporal consensus would naturally tend to result in the problematization, even pathologization, of conditions that produce temporal confusion in individuals. For functional coordination of modern society, the temporality of social systems must hold sway over the subjective time of individual psychic systems (in Niklas Luhmann's terminology). One instrument for maintaining a synchrony between individuals and social systems is autobiography. On the one hand, autobiography is the most personal and individual of histories. But its narrative articulation, whether systematically expressed in public works or not, is woven through the linear measures of modern social time, and this interpenetration is one major feature underscoring its function in the development of a particular kind of self, a modern self both opposed to and integrated with a modern society.

However, some forms of subjective experience are difficult to spread out against independent measures of social time—for example, hallucinatory experiences or lapses in memory. The temporality of psychic systems is itself dependent on the enormously complex interrelationships of neurological systems and their environments, and these sometimes run awry from social time in ways that defy acculturation. I will discuss three deliberately selected examples of this disjunction here. The phenomenon of the phantom limb, the confusions of traumatic memory, the persistence of the dead in mourning—all present forms of temporal disruption where past phenomena manifest themselves as present. And all three have been objects of a variety of modern therapeutic endeavors. However, these three situations form an unquiet series, especially in light of recent neurological advances. The understanding of phantom limbs and traumatic memory has increased significantly as a result of a huge leap in knowledge about how the brain works. But the case of mourning must be exempt from this understanding, even if, theoretically, more advanced neurological knowledge might form the basis of alleviating its pains.

Why, in theoretical terms, is mourning a particularly anomalous case in this series? The answer, I will argue, has to do with what Luhmann calls the functional latency that grounds the interpenetration of social and psychic systems.[1] Psychic systems, consisting of inner experiences, are closed with respect to each other; this closure is the basis for communication and, in turn, for the social system, precisely because it sets up the compensating necessity of relying on behavior to indicate inner experience, to orient psychic systems to each other. If the experience of each psychic system were not hidden and closed off, the social system—and communication—would be unnecessary. Yet in some respects—and this is a departure from Luhmann—behaviors do not seem to be so much the basis for communication as the source of contagious affect, a communion rather than a communication that suggests psychic system closure is not absolute but variable. Certain forms of temporal disruption offer moments where the formerly constituted closure of psychic systems shifts so that psychic systems in certain respects merge, forming a system that consists of a multiplicity of psychic systems. This shift in the identity of the system brought about through a simultaneous perception and experience of another as oneself dissolves the structural interpenetration of social and psychic systems. Phantom limbs, trauma, and grief, I argue, comprise a series in which the treatment of the particular manner of the presence of the past peculiar to each condition teeters from modern therapeutic measures intended to salvage narrative, linear autobiography for individual psychic systems to a quasi-involuntary exploitation of an opening between psychic systems. I argue, ultimately, that the periodic experience of such an opening might form a therapy for modernity itself, particularly as it develops more detailed knowledge of the neurological subsystems that produce psychic life and as it develops more abstract and global synchronies. To illustrate the point, I bring contemporary work on the brain together with the systems theory of Humberto Maturana, a foundation for Luhmann's theory of social systems, in order to consolidate a redescription of the temporal disruption of psychic systems. I then consider the work of Maturana's partner, Francisco Varela, and his colleague Natalie Depraz, together with the implications of the relatively recent discovery of mirror neurons, as challenges to the purported closure of psychic systems. I end by suggesting an important role for locality within a relatively deterritorialized global modernity.

Humberto Maturana coined the term "autopoiesis" as the result of students' questions concerning the nature of life, a persistent question in the history of biology. Individual components of living systems—their atoms and molecules—in most cases have a fleeting association with the living entity; they are constantly replenished and replaced during metabolic processes, and such systems often grow and change in the course of what is called their lives. Given this continuous flux of material elements, what is it we call a single life? What is it that persists as long as a creature retains its identity and departs at that identity's disintegration? Autopoiesis, literally "self-making," highlights the activity of living systems, not the stability or invariance of any particular component or material. This activity consists of a circular relationship with the environment; autopoietic systems are both affected by and affecting of the environment, continuously and recursively. And it is this circular pattern of activity that Maturana designates as the hallmark characteristic of life. This continuous activity both implies and requires a dynamic environment and a dynamic system; change is not merely a characteristic of autopoietic systems but a necessary trait.[2]

Maturana ascribed "cognition" to all autopoietic systems by definition: cognition is the living being's capacity to sense those features of the environment to which it responds with self-maintaining activities. Such a system's "cognitive domain" constitutes a particular specification of reality, a spectrum of salience. There may be available to an observer features of a given environment that exist outside the cognitive domain of an observed system. It may even be the case that such features are conditions for the continued integrity of the observed system. Moreover, the very specification of a "system" and an "environment" is relative to a given observer. The observer is a system, too, and the capacity to discriminate other systems and environments would depend on features of its cognitive domain. Furthermore, each system's cognitive domain is closed: it cannot share information with or transfer information to other cognitive domains that are also closed to it; it can only, through behavior, develop resonances with other systems. Finally, a primordial form of temporality is basic to all autopoietic systems as Maturana describes them. All such systems are, as he puts it, "predictive"; they have a selective relation toward the environment based on the past. The principle governing this selection is "what has happened before will recur." This principle follows from the dynamic quality of auto-poiesis itself; some stability—that is, some pattern of recurrence—must be the case for there to be minimal integrity in the autopoietic system. A completely chaotic environment could not sustain autopoiesis; by definition, no patterned recurrence of features necessary to life could exist.

The nervous system and the systems that sustain it form Maturana's chief domain of investigation. These systems' features illustrate principles of many complex systems composed of other systems, such as the system of society. The nervous system consists of neurons; it is itself a component of the system of the organism, and each of these systems' designation as such depends on an observer. And each of these systems, including the observer, has distinct cognitive domains that are not only closed to each other but differently constituted, notwithstanding their tightly coupled interdependencies. The neural cell's cognitive domain consists of electrochemical features that initiate an energy transfer to other neurons or not, depending on both the electrochemical character of the environment and the neuron's own electrochemical character at a given moment. The nervous system, made up of neurons of specific types, responds to a particular category of stimulation; the optical sensory system, for example, responds to photons; skin receptors respond to pressure, temperature, and so forth. But in addition, the organism with a nervous system is afforded a special possibility for increased freedom. Within certain threshold limits, it may modify its response to sensory input based on perusal of *internal* representations, such as memories. For this to be true, it must be the case that sensory input has the capacity to present a specific domain in which it is possible to offer a given observer options from which to select a response. Although the systems that support this possibility may be entirely deterministic—and indeed from a more remote viewpoint all systems involved may be deterministic—from the point of view of the system capable of experiencing its own internal representations, the environment offers choices.

Only such systems, unlike those that sustain it, "have" a past and not just a predictive orientation. They encounter phenomena not only through the on-going present events of the sensory system, but also through the products of their own internal processes. It is, therefore, possible for such systems to inter-rupt attention to the environment; they can attend to themselves in order to determine features that call for notice. In so doing, their cognitive domains are obviously very different from the neurological subsystems that support them. Furthermore, through the development of communication it becomes possible for one organism to orient behavior in other organisms with similar capacities.

Maturana—like Luhmann after him—insists communication is not the transfer of information. Because of the autopoietic closure of each observer's cognitive domain, there is no transfer. Instead, language has an orienting function; it works as an environmental cue to attend to a specific subset of possibilities. As such, it depends on context, without which specific utterances remain ambiguous. This ambiguity is not a shortcoming of communication. It allows a finite repertoire of signs to refer to a nearly infinite variety of circumstances. Language facilitates coordination of internal representations in external contexts that steer conduct and provide it with meaning.

For present purposes, I want to stress that the orienting function of language also has a role internal to the language user. It is part of what makes possible an autobiographical self. More generally, the relativity of viewpoint to the constitution of systems and environments permits a redescription of autobiographies that modern individuals take for granted as central to identity. Clearly, autobiography emerges from an interpenetration of psychic and social systems; individuals peruse internal representations of experience in terms acquired from social contexts, especially language. The organization of autobiography is temporal, articulated according to measures established in social systems, such as calendar and clock, even if this only serves to highlight a distinct subjective sensibility. Such measures structure the examination of memory and assist the task of ordering it. The linear projection of this order into a past that deepens into a darkness beyond the reach of memory maintains autobiographical interrelationships and confirms historical cohorts, predecessors, and successors that may not be part of direct experience at all. Overall, autobiographical identity rests on an astonishing range of interdependent temporalities: social constructions rest on psychic systems whose experienced temporality is variable, and these in turn rest on organismic systems with no time except the present—only their predictive orientations. The simple, seemingly given and natural experience of the linear time of autobiography depends on the obscurity, the inaccessibility, of the complicated weave of nested systems that enable it.

Alterations in organismic subsystems can radically affect the experience of time. This ordinarily happens nightly, when dreams manifest as external representations what are in reality internal. The absence of tonus, the inhibition of muscular movement, prevents the enaction of perceptions that in waking states provide a continual reinforcement of the distinction between internal and external representations. Conventionally, dream events are excised from autobiography—except *as* dreams. They remain, for the most part, easily distinguishable from real autobiographical events. But other events in the life of the individual also confuse distinctions between internal and external, past and present, troubling both the spatial and temporal dimensions of the subject of autobiography. Phantom limb pain is one of these, the puzzling persistence of an absent body part in the present. The persistence of sensation in the vanished limb is a long-standing historical puzzle; phantom limbs have been the object of psychophysiological inquiry and speculation from antiquity. Throughout, their presentation is described as inconvenient, often painful, and sometimes bizarre in its particular manifestations.

Hypotheses about the cause have run the gamut from purely physical to purely psychological. In the aftermath of the Civil War, the great number of amputees produced a wealth of speculation about phantom limb activity. One of the most popular notions was that the scarred end of cut nervous tissue in the stump was the cause.[3] Twentieth-century psychoanalysts even to the present day have treated the phantom as a product of melancholy, the introjection of a lost part of the self. But recently the work of neurologist V. S. Ramachandran has illustrated that an array of physiological systems contribute to the phantom limb phenomenon. From an autopoietic standpoint phantom limb pain might best be regarded as an alteration in the way a number of embodied systems interpenetrate each other, altering the cognitive domain of sensory awareness to the point that it contradicts knowledge of the body's borders.[4]

Phantom limb pain consists of false sensory reports. Even though there is no limb, the observer experiences the limb and, unless inhibited by an act of will, responds to those reports as if the limb existed. Simple denial of the sensory data does no more good than denial of valid sensory data; the phantom is not the result of insufficient conviction of the limb's absence. Ironically, as it turns out, such conviction may even help promote the phantom. Ramachandran developed a unique therapy that exploits the interpenetration of limb-related tactile data to visual data. Normally vision confirms aspects of body image available to the eye, and of course the vision of amputees normally confirms the absence of the limb. However, Ramachandran discovered that in the case of the phantom, this reinforcement operates dysfunctionally. By confirming the limb's absence (or perhaps more accurately, by failing to confirm its presence), vision can inhibit the body's updating of its own condition and thus participate in causing a failure to revise the body's map to reflect the limb's absence. Ramachandran explored this phenomenon through a deliberately constituted illusion. He invented what he called a "virtual reality box," an assemblage of mirrors that reflected and refracted the real limb, doubling it into its lost mate. Movement established connection to the image: when the patient flexed the real limb, the refracted image of the existing limb moved as well. This pattern of enaction and its visual confirmation were sufficient to establish a felt illusion that the patient was moving the phantom. This perceived experience of voluntarily activating the limb had in many cases a significant effect on the phantom's existence outside the box, generally ameliorating the extremity of sensation and in some cases eliminating the phantom all together.

Frustration at the violations of autobiographical space-time comprised by the phantom limb indicates the autopoietic closure of psychic systems with respect to organismic systems. One cannot will the phantom out of existence. The phantom limb extends the body's borders into empty space, and it offers as a perception that which autobiographical experience dictates must be only a memory. The past is that in which events have irrevocably lost their event-ness and become confined to internal representation, sometimes subject to willed recall. Organismic systems do not have a "past," but events from organismic systems bear crucially on the condition of psychic systems. And discovering the codes that determine these various systems' interpenetration allows new degrees of freedom in therapeutic assays. An observer of such systems such as Ramachandran can manipulate events on the neurological level by manipulating behavior in order to affect perception in certain patterned ways. The success of such manipulations in turn reveals the nature of perception's interaction with the new neurological terrain produced by the trauma. In this case, it emerged that visual interaction with a nonexistent limb affects whatever it is that produces tactile sensations in the phantom. The addition of visual sensory reports, even though false, initiates processes that subsequently alter the neurological basis of tactile sensations in the limb, diminishing their activity.

For psychic systems, the process features an inverted logic: visual evidence for the existence of the phantom reduces tactile evidence for the existence of the phantom. Reality for the psychic system is, in the specific case of the phantom, paradoxical; the limb is truly a phantom where the rules of space-time do not apply, and it must be exorcised through procedures that have no connection to the felt experience of navigating the world while maintaining bodily integrity. But despite the counterintuitive nature of the therapy, the amelioration of phantom limb effects brings autobiographical self into closer correspondence with messages from the body about its borders, messages only a psychic system can receive. In the process, the psychic system learns to take its own complexity into account and to use knowledge of organismic systems for functional ends, even if that knowledge prescribes acting in ways that contradict intuitions previously held as unquestionable.

Phantom limbs present psychic systems with frustrating paradoxes but do not disrupt the temporal identity offered by autobiography very deeply.[5]

Phantom limb sufferers are not brought to question the limb's nonexistence as an autobiographical fact; they do not forget its loss as an event. But the confusion of autobiographical reality is arguably more profound when the trauma more directly involves the enormously complex systems responsible for internal representations of autobiographical experience itself, of identity—that is, memory. Traumatic memories, more descriptively called "intrusive memories" in the psychological literature, are those that behave more like perceptual events than like memories and thus directly disrupt the linear time of autobiographical history, a temporality supported by the linear technologies of synchronization offered by social systems. The sufferer—that is, the experiencer of past events—has far more difficulty achieving a second-order stance than the phantom limb sufferer. In the latter, memory contradicts tactile sensation, but since the memory of the loss of the limb is not itself affected, these tactile sensations submit to a kind of special case classification that the observer can objectify and treat as a problem. But when memories themselves behave like perceptions, intruding into the present as if they were real-time events, psychic life itself must become the object of observation—seldom a possibility for the trauma sufferer but quite possible for other observers, particularly psychoanalysts and psychologists, although their observations are, in the mainstream, limited to behavior and communication of the trauma sufferer.

Like phantom limb research, trauma treatment has benefited from the view that individual experience has complex neurological origins wherein an array of systems form environments for each other. "Dual representation theory" in the cognitive study of trauma structures investigation into posttraumatic experiences in terms of conflicting sources of the internal representation of the traumatic event.[6] A major feature of posttraumatic stress disorder (PTSD) is the recurrence of unwanted memories of the traumatic scene that erupt into awareness without deliberate recall.[7] As has become a commonplace observation, trauma produces dissociation of attention; victims frequently report out-of-body experiences or an extremely narrow focus on a patch of sky or grass in place of the general scene. And experimentally, deficits in attention to the traumatic event have been convincingly correlated with the incidence and severity of posttraumatic stress. But more recently this observation has been refined: PTSD correlates with a particular division of attentional labor during the triggering event. Dual representation theorists have identified two neurological systems for processing events that encode memories in two different memory systems. The first, verbally accessible memory (VAM), is responsible for autobiographical memory, a record of experience that is context-rich and thus temporally marked. It is also more available to updating and correction by new, consciously processed experiences. Due to this kind of memory, for example, knowledge that one's mugger was caught and imprisoned can significantly decrease the sense of unease that formerly accompanied excursions out of the house. This access to new knowledge and the fact that such new knowledge affects emotional response to formerly traumatic environments reflects the neurological structures with which VAM is allied, the hippocampus and the prefrontal cortex, both relatively recent evolutionary features that play major roles in consolidating memories. This form of memory is also likely to be isomorphic with so-called declarative or explicit memory, such as the memory of last night's dinner or the capital of Oklahoma, in contrast to implicit or nondeclarative memory that consists of skills (how to ride a bicycle, how to speak).[8]

Alongside VAM and normally integrated with it, dual representation theorists have identified situationally accessible memory, or SAM. SAM seems to include implicit memory and to consist largely of perceptual phenomena in nondeclarative form. It does not project into the hippocampus and cannot be directly updated. Instead of being subject to direct declarative recollection, SAM operates in response to environmental triggering. Encounters with perceptions associated with those stored in SAM provoke a raw, uncontextualized recall, a form of memory closer to physiological reexperience of the original event than to the memories we consciously peruse and manipulate. Not surprisingly, SAM is neurologically older. It has a close relationship to the amygdala, the brain structure associated with negative survival-related emotions such as fear and anger.

These two systems with their different forms of recall relate to different kinds of encoding data, VAM narrative and verbal, SAM perceptual and often predominantly visual. This division of labor has been the basis for a number of experiments that attempt to determine the way traumatic events are encoded and to exploit these differences therapeutically. Recently researchers confirmed evidence that tasks drawing on SAM, visuospatial tasks such as tapping a keyboard during the experience of traumatic material (a controlled situation such as a film of a car crash) significantly reduce the likelihood of subsequent traumatic intrusions. Furthermore, verbal distraction during the experience of trauma—a distracting conversation or verbal memory task—increases the likelihood of subsequent intrusions. Researchers explain that activities loading SAM during trauma encoding reduce this memory system's responsiveness to trauma perceptions, and as a result there may be greater activation of VAM; contrarily, loading VAM with verbal distraction, perhaps not only with conversation but with dissociative daydreaming, impoverishes the VAM encoding of traumatic events and might enhance the role of SAM. Because SAM memories have impoverished context and context markers serve to fix particular memories temporally, the greater role SAM has in encoding events, the more subject those event-memories are to involuntary reemergence through an array of environmental cues. Furthermore, researchers were able to affect traumatic memory through the exploitation of VAM and SAM not only during the original encoding, but also during episodes of traumatic recall after the triggering event.

This nuanced, multiple-system approach to memory heavily qualifies the idea that traumatic memory intrusions and other pathological markers of PTSD result from deliberate repression. To be sure, the effects of "mental control" of traumatic experiences have been experimentally shown to correlate with an upturn in traumatic intrusions. However, it would now appear that this correlation may not be due to repression itself, as in a Freudian model, so much as in the increased demand on VAM that repression requires, a competing narrative intended to snuff out the narrative of traumatic experience.[9] Thus it is more likely to be the case that memory therapies as treatment of PTSD are not so much encountering resistance toward the memory so much as an impoverishment of the narrative form at the time of encoding. Furthermore, there have been indications that inducing nonverbal perceptual tasks during PTSD treatment significantly after the encoding has more efficacy in reducing traumatic symptoms than talk therapy.

No one would wish PTSD sufferers to endure their affliction with any more extremity or duration than necessary. At the same time, however, dual representation theory's exploitation of a second-order relationship of observation to memory has more complicated implications in the domain of psychic systems than it had in the case of the phantom limb; the mechanical relief of anguish by means of seemingly unrelated tasks is somewhat unsettling. Like the ironies of Ramachandran's virtual reality box, visual-spatial tapping tasks seem a strange answer to the form of suffering—and in this case there is an additional affective dimension: the observer of traumatic experience in another often has an empathic response. In fact, trauma on a large scale—the Holocaust, the collapse of the Word Trade Center towers—has been the occasion for a form of social suffering that has itself been deemed healing by some writers. Such healing utilizes the sympathetic contagion of traumatic affects, dependent on the heightened emotional excitation of the trauma sufferer, to resocialize the trauma. This sympathetic contagion in turn rests on the continued latency of the multiple neurological components of traumatic memory.[10] In other words, the tangle of autobiographical time produced by the disrupted integrity of VAM and SAM memory systems itself has a social effect; it appears to assist constituting an affective bond between the trauma sufferer and those who listen and observe. To undo the latency of the dual system memory endangers this integration of psychic and social systems in this respect.

Yet it may be incorrect to characterize this social effect as truly social in Luhmann's sense. At least as it is ordinarily discussed, empathy violates the premise of psychic system closure. The experience of empathy is the experience of another's psychic condition as in some way one's own. Of course, this apparent communion of psychic systems has frequently been explained as imaginative projection or as the unthematized triggering of one's own memories through the behavior of another. However, the discovery of mirror neurons in the primate cortex suggests that this explanation may be misguided; in a truly fundamental sense, the body of another can be experienced as one's own without the mediation of imagination.

Researchers discovered mirror neurons, a class of motor neurons, in the early 1990s.[11] They noted that a peculiar group of neurons in the ventral premotor cortex of monkeys fired not only upon the initiation of the monkey's own motor activity; they also fired when the monkey observed deliberate motor activity in another monkey, an activity such as reaching for a banana. In other words, these neurons fire both when the action belongs to the home body and when the action is that of another body. In the succinct words of Cristina Becchio and Cesare Bertone, these neurons "match the sound or vision of *someone else's* actions onto the *subject's own* repertoire."[12] So the other body's viewed or heard behavior is, with respect to certain activities, one's own body's behavior at the neurological level; there is, truly, shared experience, in the sense that for the mirroring monkey there is no other monkey at this stage of cognition.

While the degree to which mirror neurons account for intersubjective experiences often described as sympathetic or empathetic remains to be determined, it does appear, as Becchio and Bertone argue, that they profoundly challenge the thesis of psychic closure that has been a cornerstone of epistemology and psychology since Descartes. The effect of such neurological interidentification for autobiography is, to say the least, disruptive. In the case of the phantom limb, the effects of observing the limbs of others reaching and grasping when one's own corresponding limb is missing might be part of the reason Ramachandran's visual restoration of the limb in the virtual reality box works. But in a sense, to the degree that mirror neurons might account for this, what the virtual reality box restores is not so much a consolidated autobiography as a heterobiography. Furthermore, more generally, the motor/sensory duality of mirror neurons maps onto the activity/passivity duality that often appears at the core of phenomenological discussions of affect and temporality. Francisco Varela and Natalie Depraz have argued that the origin of temporality that constitutes the infrastructure of narrative and social time lies in a microtemporality situated at the transitional point between prereflexive and reflexive experience, a polar self-other in which self is other and other is self, a multiplicity in a state of dynamic tension that precedes second-order cognition. "This is the manifestation . . . of the inseparable presence of passivity and agency," write Varela and Depraz.[13] Given this analysis, the closure of psychic systems with respect to each other succeeds rather than precedes a primordial shared temporality consisting of embodied experience, an instance of self-affection that is indistinguishable from other-affection. Conversely, social communication rests on communion's extinction, brought about by reflection and psychic closure. But that closure also brings with it a "waning of affect," to use Fredric Jameson's phrase, a transformation of the polar tension comprising the multiplicity/unity of self-and-other into the relatively static duality of subject and object, self and other.[14]

Other bodies witnessed or heard trigger the state of anti-Cartesian communion—and this understanding might help explain the compelling embodied quality of mourning. The witnessing of mourners in deep grief, even by strangers, often produces a profoundly powerful empathy. And grief is a physical experience for the mourner. The loss of an intimate, a lover or close family member, leaves in its wake a variety of bodily movements whose superfluity, like that of feeling in a phantom limb, brings a sudden halting ache, a condition famously portrayed in William Wordsworth's sonnet, "Surprised by Joy":

> Surprised by joy—impatient as the wind
> I wished to share the transport—oh, with whom
> But thee, long buried in the silent tomb,
> That spot which no vicissitude can find?

This kind of habitual inclination toward the dead, not so much described as indicated by Wordsworth's language, summons an involuntary sympathetic pause in the reader. It is like the automatic reach across an empty bed, the dialing of a phone number that will connect to no one, that painfully frays the autobiographical fixation of the death at a point in the past.[15] These embodied habits whose performance is so affectively contagious are detemporalized and do not disappear in synchrony with the linear narrative of life and death. On the one hand, the trouble posed by the dead to autobiography may be captured and confined by convention, but, on the other hand, the famous black page in *Tristram Shandy* mourning the death of Parson Yorick signifies the impossibility of psychic system closure in this context. Even if these painful automatic movements toward and with the absent figure might be banished through the exploitation of the VAM/SAM distinction, it is far from certain this would be a popular therapy, and this is not merely a matter of propriety. Rather, it concerns the bond of embodiment mediated by objects of desire in the visible, audible world, objects that, even when absent, evoke a multisubjective grasp and reach, a simultaneous action and perception, activity and passivity.

The state of mourning is probably heavily involved with SAM, stimulated by the recurrence of associated environmental features, by bodily habit and automaticity. Rich in perceptions, tightly connected to emotional arousal, the recurrence of the dead in bodily acts is likely to be diminished by eliminating familiar context, by change of place, by modern geographical mobility. The bodily manifestation of motive and emotion underlying intersubjective communion relies on sensory connection to the physical environment. Moderns leave the triggers of perceptual associations behind or tear them down in fits of renovation and gentrification. Wordsworth, clearly the poet of SAM, lamented the loss of ancestral lands as a cause of profound psychic and economic dislocation in the advent of the Industrial Revolution.[16] The significance of locale in Wordsworth's sense is inseparable from its character as the site of long-term communal sharing. The connection of the dead to a particular place, the fact that "haunting" requires a particular location physically connected to the dead, is often poignantly illustrated by the complaints of displaced peoples. To cite just one example, one of the world's most ancient cultures, that of the Kalahari Bushmen, is fast disappearing through the tribespeople's relocation into camps and towns. A *Washington Post* article recently depicted the forced transfer of a village by the government of Botswana, ostensibly to replace the Bushmen's hunting-dependent livelihood with agricultural practices more compatible with territorial confinement. As one woman reports, the ancestors communicate their displeasure in dreams. "You have lost us," her deceased grandparents told her. "Why are you not next to us?"[17] The grandparents' dream appearance is clearly a poor substitute for their felt presence in a familiar physical environment.

Dual representation theory suggests that the strong attachment of traditional culture to the dead is more than a social construction; it arises from the physical nature of embodied memory, strongly activated by associatively charged environmental cues. When environments change, these triggers are lost and the vivid feeling of the presence of the past goes with it. Geographical displacement may diminish the intrusive effects of the past without explicitly manipulating memory systems as such, and it has been a long-standing feature of modernity's global progress. But from a second-order point of view, other questions must be raised: to what degree does the persisting body of locale anchor the connection between the bodies of the social beings that occupy it?

Might modern mobility weaken the role of SAM or of affect generally? Perhaps a threat to affective communion partly accounts for recent preoccupation with memorializing trauma, from Holocaust museums to AIDS quilts to anonymous roadside flowers. Perhaps, inexorably, the necessity of restoring conditions for intersubjective experience exerts a counterpressure to the demands to erase it or revise it, and there may be more a growing resistance to modern displacement and the rise of movements aimed at helping to secure persons against a corrosive functionalism to the benefit of both psychic and social systems.[18] In any case, it would appear that the closure of psychic systems foundational to Luhmann's conception of social systems rests on a condition that is deeply ambiguous with respect to closure. Both other beings and events, in at least some respects, participate in an intersubjective microtemporality that suggests new grounds for interrogating the functional latency upon which modernity depends.

Notes

1. Luhmann, "Temporalization of Complexity," 100–101.
2. The description of autopoiesis that follows is a summary of sections 3 and 4 of Maturana's *Biology of Cognition*, as reprinted in Maturana and Varela, *Autopoiesis and Cognition*, 8–40.
3. Ramachandran and Blakeslee, *Phantoms in the Brain*, 23.
4. Ibid., 50–56.
5. Damasio deemed the phantom limb "tolerable in the long run." See his *Looking for Spinoza*, 193.
6. Brewin, *Posttraumatic Stress Disorder*, 104–27.
7. This signature recall experience is the basis for most recent explorations of dual representation theory. See Holmes, Brewin, and Hennessy, "Trauma Films," 3.
8. Brewin, *Posttraumatic Stress Disorder*, 119–22.
9. This interpretation is much more consistent with that of Pierre Janet, who anticipated dual representation theory. See Holmes, Brewin, and Hennessy, "Trauma Films," 18, and Brewin, *Posttraumatic Stress Disorder*, 108.
10. For a volume of essays that frequently pursues the social value of witness to trauma, see Caruth, *Trauma*.
11. For a clear, summary article on mirror neuron research and implications, see Ramachandran, "Mirror Neurons and Imitation Learning."
12. Becchio and Bertone, "Beyond Cartesian Subjectivism," 23; emphasis in original.
13. Varela and Depraz, "At the Source of Time," 75.
14. Jameson, *Postmodernism*, 10–16.
15. For a powerful exploration of mourning as embodied experience, see Krasner, "Doubtful Arms and Phantom Limbs," 218–32.

16. In a letter of 1801, Wordsworth describes the power of such land in the lives of its rustic proprietors to the great Whig statesman Charles James Fox: "Their little tract of land serves as a kind of permanent rallying point for their domestic feelings, as a tablet upon which they are written which makes them objects of memory in a thousand instances when they would otherwise be forgotten. It is a fountain fitted to the nature of social man from which supplies of affection, as pure as his heart was intended for, are daily drawn. This class of men is rapidly disappearing" (Wordsworth, *Letters*, 1:314–15).

17. Cited in Timberg, "A Culture Vanishes in Kalahari Dust."

18. For a survey of various systems-theoretical discussions of the value of desynchronization, see Brose, "An Introduction towards a Culture of Non-Simultaneity."

Meaning as Event-Machine, or Systems Theory
and "The Reconstruction of Deconstruction"

Derrida and Luhmann

CARY WOLFE

As I said: humans can't communicate.
—NIKLAS LUHMANN, "How can the mind participate in communication?"

The reception of systems theory in the United States—and in North America generally—over the past decade and more has been vexatious at best. In a professional academic landscape in which most critics and theorists pride themselves on moving easily and syncretically between theoretical approaches that, at an earlier moment, were thought of more as warring factions, systems theory remains odd man out. When it is understood at all, it is routinely greeted with reactions ranging from suspicion to outright anger. Critics who think of their work (rightly or wrongly) as a component of a broader political project—at least "in the last instance," to borrow Louis Althusser's well-known caveat—often view systems theory as just a grim technocratic functionalism or a thinly disguised apology for the status quo, a kind of thinly camouflaged social Darwinism. On this view, systems theory—in either its first-order, Norbert Wiener version or its second-order, Niklas Luhmann retooling—gets assimilated to the larger context of post–Second World War society's obsession with management, command-and-control apparatuses, informatic reproduction, homeostasis, and the like, rightly criticized by theorists like Donna Haraway in her important essay, "The Biological Enterprise: Sex, Mind, and Profit from Human Engineering to Sociobiology."[1] Systems theory, instead of being invited to the party reserved for chaos and complexity theory and their interest in the unpredictability, creativity, and emergence of complex nonlinear dynamics, ends up dancing with Richard Dawkins's *The Selfish Gene*. Still others level more general charges familiar from the shopworn discourse of anti-theory and lament systems theory's excessive abstraction, its lack of attention to social and historical texture, and its blind ambition to assimilate everything in its purview as grist for its universalizing mill.

If these charges sound familiar, they ought to, because they are an uncanny echo of the sorts of things that we all remember being said about deconstruction—and specifically about the work of Jacques Derrida—when it came ashore in North America in the 1970s (Derrida's *Speech and Phenomena* appeared in translation in 1973, followed in rapid succession by *Of Grammatology* in English in 1976 and *Writing and Difference* in 1978). Of course, we all got over it, and the irony need hardly be remarked (but I'll remark it anyway) that it is difficult to find people who have had much success in the profession of literary and cultural studies in North America who did not cut their teeth on just these texts and whose deployment of lessons learned from them in their own work is not more or less automatic and unconscious.

The reasons for systems theory's chilly reception in the United States are complicated, I think, and I'm not going to investigate them in any detail here, but I'll at least offer a couple of very brief speculations. One set of reasons (not to be underestimated) is disciplinary and institutional. First, as many of us remember, "deconstruction in America"—a time capsule phrase if ever there was one—made its way into universities mainly via comparative literature departments; and if you think *that* was a precarious foothold, consider the fact that the major practitioner of systems theory (Luhmann) has entered the U.S. academy primarily by way of *German* departments (or their equivalent fractions in larger comparative literature and language departments), mainly under the rubric of German intellectual history. (Here the work of scholars such as David Wellbery, William Rasch, and Hans Ulrich Gumbrecht would be exemplary.) But over the past decade, many American universities have downsized or eliminated altogether their German departments, and it is hard for me to think of any more endangered place to be in the humanities in the United States over the past ten years, with the possible exception of classics departments.

Related to this question of institutional foothold is another, different deficit: the absence of a nationally disseminated journal that is tethered more or less unilaterally to the theoretical model. *Diacritics* (published out of the Comparative Literature Department at Cornell University) became something like the house journal for deconstruction in the 1970s and '80s, but the *Stanford Literature Review*, which has done more than any single U.S. journal to consistently publish work in systems theory, is not *Diacritics*. A few special issues of other, well-known journals have been devoted to systems theory and/or Luhmann's work—*Theory, Culture, and Society* (published by Sage in

Great Britain, though widely available in the United States), MLN, *New German Critique*, and one and a half issues of *Cultural Critique* entitled "The Politics of Systems and Environments"—but nothing that has the kind of ongoing relationship to systems theory that *Representations* did and does for New Historicism. Moreover, systems theory has had to brook an even greater degree of disciplinary dissonance; where the establishing texts of deconstruction were quite identifiably within the purview of philosophy and often of literature, the major texts and figures of systems theory enter the humanities through the side door of *science*—either social science and sociology (with Luhmann) or the life sciences (in the case of Humberto Maturana and Francisco Varela) or the interface of first-order cybernetic computer science with neurology (in the case of Heinz von Foerster). And finally, of course, there is the daunting difficulty of the theory itself, which—particularly in Luhmann's hands—gives even seasoned readers of theory pause with its extraordinary abstraction and rigor; its head-on engagement with problems of paradox, self-reference, and the like; its systematically counterintuitive findings; and its relative lack of creature comforts along the way for those who have signed on for the journey of what Luhmann unabashedly calls "super-theory." Of course, here again we should probably remind ourselves that it is hard to recall a major theoretical development about which something similar was *not* said, and some of our colleagues are old enough to remember similar complaints about the technical rigor and cold-bloodedness of that strange, alienating, scientistic approach to literary texts called "the New Criticism."

Other speculations could no doubt be offered about why systems theory in the United States has not emerged as the kind of factor in cultural studies that it is most obviously in Germany, but my main point is not to analyze those reasons further. Rather, it is to nudge the reception of systems theory in a different direction by strategically bringing out some of its more "deconstructive" (if you will) characteristics.[2] Indeed, I hope to make it clear to skeptics that much of what they like about deconstruction is also much of what they *should* like about systems theory because systems theory in its contemporary articulation—far from conforming to the stereotypes prepared for it in the U.S. academy—"may well be read," to borrow Dirk Baecker's formulation, "as an attempt to do away with any usual notion of system, the theory in a way being the deconstruction of its central term."[3] To take only one example, let us revisit the epigraph with which I began. On the one hand—the dominant hand—Luhmann's contention that "humans can't communicate" seems not just counterintuitive but flatly wrong; in fact (as a colleague mentioned to me recently at a conference), it seems "insulting."

And yet—as will be clear by the end of my comments here, I hope—Luhmann's remark (rhetorically calculated, no doubt, to cause just such a stir) makes essentially the same point about the difference between "consciousness" and "communication" that we have quite readily accepted for decades now as gospel from Derrida—namely, his deconstruction of the "auto-affection" of the voice-as-presence and of the valorizing of speech (as an index of the self-presence of consciousness to itself) over writing (a recursive domain of iterative communication that is, properly understood, fundamentally ahuman or even anti-human).[4] Indeed, as we will see in more detail below, it is to insist on just this sort of radical separation of what Luhmann calls psychic and social systems that Derrida comes to reject the notion of "the signifier" (as in Lacan's formulation of "the subject of the signifier," which seems at first glance quite cognate to Luhmann's formulation) in favor of the articulation of writing as fundamentally a structured dynamics of "the trace."

My pairing of systems theory and deconstruction here should come as no surprise because Derrida himself announces the convergence in his own terms in early, formative texts such as *Of Grammatology*, whose first chapter, "The End of the Book and the Beginning of Writing," begins with a section entitled "The Program." There Derrida argues that "the entire field covered by the cybernetic *program* will be the field of writing," but writing understood in terms of "the *grammé*—or the *grapheme*," a writing that would name as its fundamental unit "an element without simplicity"—which is to say, as I will argue below, an element of irreducible *complexity* (specifically as systems theory uses the term). And a temporalized complexity at that, for as Derrida argues, "Cybernetics is itself intelligible only in terms of a history of the possibilities of the trace as the unity of a double movement of protention and retention."[5]

Derrida's claim, forwarded as it was in the late 1960s, may seem even now a radical one, but in fact it was lodged against the backdrop of an entire revolution in the sciences that had already taken such models as axiomatic. In fact, the first chapter of 1965 Nobel Prize winner François Jacob's remarkably influential *The Logic of Life* is *also* called "The Programme." There Jacob reminds us that "heredity is described today in terms of information, messages, and code," and what this means—and this is clearly related to Derrida's early work on both Husserl and Saussure—is that "the intention of a psyche has been replaced by the translation of a message. The living being does indeed represent the execution of a plan, but not one conceived in any mind."[6] Derrida would add to this, however, the point he presses in *Of Grammatology*: "If the theory of cybernetics is by itself to oust all metaphysical concepts—including the concepts of soul, of life, of value, of choice, of memory—which until recently served to separate the machine from man, it must conserve the notion of writing, trace, grammé, [written mark], or grapheme, until its own historico-metaphysical character is exposed."[7] As an example of such "character," interestingly enough, Derrida cites in a footnote not Jacob but first-generation systems theorist Norbert Wiener, who "while abandoning 'semantics,' and the opposition, judged by him as too crude and too general, between animate and inanimate etc., nevertheless continues to use expressions like 'organs of sense,' 'motor organs,' etc. to qualify the parts of the machine."[8] Part of what I will be arguing in what follows is that Luhmann's handling of systems theory accomplishes just the sort of "conservation" of the logic of the grammé that Derrida calls for, a conservation that is crucial to any posthumanism whatsoever—not only because the movement of the program-as-grammé "goes far beyond the possibilities of the 'intentional consciousness'" as the source and guarantor of meaning, but also because once the notion of the program is invoked, one no longer has "recourse to the concepts that habitually serve to distinguish man from other living beings (instinct and intelligence, absence or presence of speech, of society, of economy, etc.)."[9]

Now as I have suggested elsewhere, this cross-talk between postwar science and what would come to be called "theory" is not limited to Derrida and Wiener; indeed, perhaps the most profound backstory of all in contemporary thought is the ongoing, if episodic, influence of such new scientific discourses upon thinkers who would emerge in the 1950s and '60s to redefine the very landscape of the humanities and social sciences (think here of Foucault's interest in Jacob and Canguilhem, Lacan's in cybernetics, Lyotard's in chaos and catastrophe theory, and so on).[10] My aim at the moment, however, is not to make that historical argument. Nor is it just to play up the deconstructive aspects of systems theory, nor even to suggest, as I have been, that the largely knee-jerk reactions to systems theory in the United States have been misplaced (or at least, vis-à-vis the reception of deconstruction, rather ungenerously placed).

Rather, my emphasis here will be on the usefulness of viewing second-order systems theory as (to use Niklas Luhmann's characterization) "the reconstruction of deconstruction."[11] That project hinges on systems theory's extraordinarily rigorous and detailed account of the fundamental dynamics and complexities of meaning that subtend the reproduction and interpenetration of psychic and social systems. And systems theory then takes the additional step of linking those dynamics to their biological, social, and historical conditions of emergence and transformation, a crucial move that, as Gunther Teubner has argued, deconstruction either cannot or will not undertake. It is certainly the case that Derrida's later work has been intensely and increasingly engaged with the question of social institutions in all their forms—the law, the university, the question of rights, the institution of property, and so on—and the logics that ground and sustain their reproduction. But even though he has raised such questions—*worried* them might be a better term—with a degree of nuance and suppleness perhaps unmatched in contemporary theory and philosophy, Derrida has not been especially interested in articulating the relationship between the theoretical complexities of those dynamics and the historical and sociological conditions of their emergence—conditions that he himself suggests impel such thinking at this very moment.[12] (Whether this is a failure or a principled refusal on Derrida's part is an interesting question, and it is one I will return to in my remarks below.)

One could cite any number of Derrida's texts in this connection, but the recent collection of essays *Without Alibi* exemplifies quite well what I mean. There Derrida considers the question of what he calls "a politics *of the* virtual," of "a certain delocalizing virtualization of the space of communication, discussion, publication, archivization," against the backdrop of this larger question: "Will we one day be able, and in a single gesture, to join the thinking of the event to the thinking of the machine?"[13] "Today," he continues, "they appear to us be antinomic. . . . An event worthy of the name ought not, so we think, to give in or be reduced to repetition," but rather "ought above all to *happen* to someone, some living being who is thus *affected* by it." The machine, on the contrary, is destined "to reproduce impassively, imperceptibly, without organ or organicity, received commands;" it obeys "a calculable program without affect or auto-affection."[14]

If we are to address the sorts of questions raised here, Derrida argues, now is the time for a new kind of thinking. "How," Derrida asks, "is one to reconcile, *on the one hand*, a thinking of the event, which I propose withdrawing, despite the apparent paradox, from an ontology or a metaphysics of presence . . . and, *on the other hand*, a certain concept of machineness [*machinalité*]?"[15] This, Derrida rightly observes, is "the place of a thinking that ought to be devoted to the virtualization of the event by the machine, to a virtuality that, in exceeding the philosophical determination of the possibility of the possible . . . exceeds by the same token the classical opposition of the possible and the impossible."[16] "If one day," Derrida continues, "with one and the same concept, these two incompatible concepts, the event and the machine, were to be thought together, you can bet that *not only*. . . will one have produced a new logic, an unheard-of conceptual form. In truth against the background and horizon of our present possibilities, this new figure would resemble a monster."[17] It would be, in a word, posthumanist.

What I want to suggest, of course, is that systems theory in its second-order incarnation is just such a "monster," one whose cornerstone genetic mutation is the transfer of the concept of autopoiesis from organicity to the domain of not only psychic but also social systems, systems whose fundamental elements are not people or groups but communications and "events"—and events quite rigorously conceptualized along the lines Derrida lays out above. We have already used deconstruction to help clarify a central point from systems theory—the separation of psychic and social systems—but here we can return the favor and use systems theory to clarify how the thinking of the event may be, in Derrida's words, withdrawn from "an ontology or metaphysics of presence."[18] On the one hand, events constitute the fundamental elements of psychic and social systems in Luhmann's scheme. On the other hand, "they occur only once and only in the briefest period necessary for their appearance (the 'specious present'). They are identified by this temporal appearance and cannot be repeated."[19] But "precisely this suits them to be the elementary units of processes" because "the system itself determines the length of time during which an element is treated as a unity that cannot be further dissolved; that period has a conferred, not an ontological character."[20] An element's unity "corresponds to no unity in the substrate; it is created by the system that uses them through their connectivity"; "accordingly," Luhmann continues, "an adequately stable system is composed of unstable elements. It owes its stability to itself, not to its elements; it constructs itself upon a foundation that is entirely not 'there,' and this is precisely the sense in which it is autopoietic."[21] And here, as much as anywhere, we get a very specific sense of how systems theory thinks Derrida's event and machine all at once, as a deconstructive enfolding of the difference between the system's iterative self-reference and the fleeting temporality of the event from the "outside"—a difference that not only serves as the very basis for the system's autopoiesis, but also clarifies the fact, as Dietrich Schwanitz puts it, that "systems theory is anything but mechanistic."[22]

As for Derrida's part—you will have already guessed by my use of the term "iterative" above—we know what his version of this monstrosity of the event-machine looks like: it looks like *écriture*, arché-writing as *différance*, as *grammé* and as trace. For our purposes, it is all the more interesting, then, that in contrast to his notion of writing, Derrida has *interrogated* the concept of communication in a variety of contexts and nowhere more forcefully, perhaps, than in his essay "Signature Event Context" and its related documents collected in *Limited Inc.* There Derrida argues that his concept of writing can "no longer be comprehensible in terms of communication, at least in the limited sense of a transmission of meaning. Inversely, it is within the general domain of writing, defined in this way, that the effects of semantic communication can be determined as effects that are particular, secondary, inscribed, and supplementary."[23]

The full resonance of this last assertion in relation to the dynamics of "meaning" in systems theory will become clear below, but for now we need to note as well that the difference between writing in the Derridean sense and communication as he defines it is marked by radically different relations to the question of the subject—and here, indeed, we encounter, from the point of view of humanism, part of its "monstrosity." As Derrida writes, "Imagine a writing whose code would be so idiomatic as to be established and known, as a secret cipher, by only two 'subjects'"—and "subjects" here is given in quotation marks:

> Could we maintain that, following the death of the receiver, or even of both partners, the mark left by one of them is still a writing? Yes, to the extent that, organized by a code, even an unknown and nonlinguistic one, it is constituted in its identity as a mark by its iterability, in the absence of such and such a person, and hence ultimately of every empirically determined "subject." ... The possibility of repeating and thus of identifying the marks is implicit in every code, making it into a network [*une grille*] that is communicable, transmittable, decipherable, iterable for a third, and hence for every possible user in general. To be what it is, all writing must, therefore, be capable of functioning in the radical absence of every empirically determined receiver in general.[24]

Herein lies the radically posthumanist dimension of writing-as-difference: the subject—in a process nearly proverbial for postmodern thought from Derrida to Lacan—comes to be only by conforming to a strictly diacritical system of differences, "effects which do not find their cause in a subject or a substance, in a thing in general, a being that is somewhere present, thereby eluding the play of *difference*."[25] Moreover, those effects and relations are at once material, bodily, external, institutional, technological, and historical—they exist in all the specificity and heterogeneity of what Derrida will call their "iteration." Hence, Derrida will argue that "this pure difference, which constitutes the self-presence of the living present, introduces into self-presence from the beginning all the impurity putatively excluded from it. The living present springs forth out of its nonidentity with itself and from the possibility of a retentional trace. It is always already a trace." And what this means, in turn, is that "the trace is the intimate relation of the living present to its outside, the opening to exteriority in general."[26]

From the point of view of the philosophical tradition Derrida is concerned to deconstruct, such will be the "corrupting" and "contaminating" work (the "monstrosity," if you will) of "iterability," which "entails the necessity of thinking *at once* both the rule and the event, concept and singularity"; as such, it "marks the essential and ideal limit of all pure idealization," *not* as "the concept of nonideality," as ideality's pure other, but as the impossibility (or at the very least the provisionality) of idealization as such.[27] Like the undecidability that it unavoidably generates—and this will lead us to the final question we want to raise—iterability "remains *heterogeneous*" to, rather than simply opposed to, the order of the ideal, the calculable, the pure, and so on. As such, it names a form of ethical responsibility that entails vigilant attention to each specific, interfolded iteration of "rule and event," to "*this particular* undecidable" that "opens the field of decision or decidability," one that "is always a *determinate* oscillation between possibilities" that takes place "in strictly *defined* situations (for example, discursive—syntactical or rhetorical—but also political, ethical, etc.). They are *pragmatically* determined."[28]

Exactly what the force and scope of this last assertion—"pragmatically determined" and in "strictly defined situations"—is for Derrida is a question that goes directly to Teubner's concerns already touched on above, and it is one to which I will return below. For now, however, I simply want to make the point that this picture of writing in the Derridean sense (and the restructuring of the question of the subject that it pulls in its wake) does not mark a *difference* between Derrida's *écriture* and the concept of communication in systems theory; rather, it is precisely what illuminates their convergence. For when Derrida uses the term "communication" in *Limited Inc.*, what he really has in mind is the model of communication mobilized by *first*-generation systems theory. That model, like the speech act theory of Austin deconstructed in *Limited Inc.*, seems, but *only* seems (as it turns out), to rightly refer the question of meaning to its external formal dynamics rather than to ontology, intentionality, and so on. Of course, it is this very baggage attached to the term "communication" that Luhmann's work, like Derrida's, is dead set on rejecting. And in fact the chapter in *Social Systems* that makes this clearest—entitled "Communication and Action"—explicitly references Derrida's critique of Husserl in protesting that "the metaphor of 'transmission'"—the metaphor that dominates first-wave systems theory—"is unusable because it implies too much ontology" in the picture it gives of both meaning (the "message") and the subject who is part of its circuit.[29] Over and against this, as Schwanitz points out, both deconstruction and Luhmann's systems theory "make difference their basic category, both temporalize difference and reconstruct meaning as a temporally organized context of displacement and deferral. Both regard their fundamental operation (i.e. writing or communication, respectively) as an independent process that constitutes the subject rather than lets itself be constituted by it."[30]

Here, however, we find a diametrically reversed orientation or angle of approach in the two theories—one that, I believe, accounts for the "monstrosity" of deconstruction being relatively well received, while systems theory has tended to provoke all sorts of defensive and recuperative responses. To put this very schematically, Derrida and Luhmann approach many of the same questions and articulate many of the same formal dynamics of meaning (as self-reference, iterability, recursivity, and so on), but they do so from diametrically opposed directions. As Schwanitz has pointed out, the starting point for systems theory is the question of what makes order possible and how highly organized complexity, which is highly improbable, comes into being at all. Deconstruction, on the other hand, begins with taken-for-granted intransigent structures of logocentrism and the metaphysics of presence that are already ensconced in textual and institutional form and then asks how the subversion of those structures by their own elements can be revealed.[31]

For Derrida, contingency, temporality, the event, "noise," and so on constitute the eruptive and finally irrepressible difference at the heart of any logos or law, a difference whose unavoidability and unmasterability deconstruction aims to bring to light and sustain. For systems theory, however, this radical heterogeneity is handled within an adaptive and operational framework, as a fundamental evolutionary *problem* for autopoietic systems that have to reproduce themselves in the face of this overwhelming difference.[32] Because of this reversal of orientation, the *descriptions* offered by systems theory ("autopoietic systems that can reduce environmental difference and complexity will continue to exist") have been misunderstood as *prescriptions* ("such systems *should* exist and difference and complexity are negative values"). But of course, systems theory doesn't *desire* the reduction of difference and complexity (indeed, Luhmann would be the first to insist that such would constitute a category mistake if ever there was one); it only *describes* how difference and complexity have to be handled by systems that hope to continue their autopoiesis.

Systems theory, in other words, doesn't occlude, deny, or otherwise devalue difference but rather *begins* with difference—namely, the cornerstone postulate of the difference between system and environment—and the corollary assumption that the environment of any system is always already of overwhelmingly greater complexity than the system itself. Since it is obviously impossible for any system to establish point-for-point correspondences between itself and its environment, systems thus handle the problem of overwhelming environmental complexity by reducing it in terms of the selectivity made available by the system's self-referential code; as Luhmann puts it, "The system's inferiority in complexity must be counter-balanced by strategies of selection." "Complexity, in this sense," he continues, "means being forced to select," and thus, in Luhmann's winning formulation, "only complexity can reduce complexity."[33] Under pressure to adapt to a complex and changing environment, systems increase their selectivity—they make their environmental filters more finely woven, if you like—by building up their own internal complexity by means of self-referential closure and the reentry of the system/environment distinction within the system itself in a process of internal differentiation.[34]

For example, the difference between the legal system and its environment is reintroduced in the legal system itself, which now serves as the environment for the various subsystems of the law, and the same could be said, within the educational system, about the various academic disciplines and subdisciplines, and so on.[35] This self-referential closure, however, "does not contradict the system's *openness to the environment*. Instead, in the self-referential mode of operation, closure is a form of broadening possible environmental contacts; closure increases, by constituting elements more capable of being determined, the complexity of the environment that is possible for the system." And this is why, Luhmann writes, "self-reference is in itself nothing bad, forbidden, or to be avoided"; indeed, it "points directly to system formation" because systems "can become complex only if they succeed in solving this problem and thus in de-paradoxicalizing themselves."[36]

What makes such systems paradoxical in the first place—the generative self-reference that gets the ball rolling, as it were—is the unity of the difference between the two sides of the distinction that anchors its code. For example, the first-order distinction between legal and illegal in the legal system is itself a product of the code's own self-reference—that is, the problem is that *both* sides of the distinction are instantiated by *one* side of the distinction (namely, the legal; hence the tautology, "legal is legal"). But the tautological *unity* of this distinction may be disclosed only by a *second*-order observer, operating within *another* system and *another* code, which must remain blind to *its* paradoxical distinction if it is to use that distinction to process events for the system's autopoiesis, and so on and so forth.

What is most interesting here, however, is that these constitutive paradoxes, far from hindering the autopoiesis of self-referential systems, in fact *force* their autopoiesis.[37] And here—in this transvaluation of the paradoxes of self-reference from paralytic to productive—the lines of relation between systems theory and deconstruction come quite clearly into view. Luhmann writes as follows:

> If we want to observe paradoxical communications as deframing and re-framing, deconstructing and reconstructing operations, we need a concept

of meaning . . . as the simultaneous presentation . . . of actuality and possibility. . . . The distinction actual/possible is a form that "re-enters" itself.
On one side of the distinction, the actual, the distinction actual/possible
reappears; it is copied into itself. . . . If we observe such a re-entry, we see a
paradox. The re-entering distinction is the same, and it is not the same. But
the paradox does not prevent the operations of the system. On the contrary,
it is the condition of their possibility.[38]

This is so, Luhmann writes, because "The totality of the references presented
by any meaningfully intended object offers more to hand than can in fact be
actualized at any moment. Thus the form of meaning, through its referential
structure, *forces* the next step, to *selection*."[39] But that selection, of course, immediately begins to deteriorate in usefulness under pressure of the temporal
flow of events, the "specious present," which then forces *another* selection, and
so on and so forth.

Here we encounter systems theory's version of what Derrida calls the dynamic force of *différance* as "temporization" and "spacing," as "protention" and
"retention," a process that "is possible only if each so-called present element . . .
is related to something other than itself, thereby keeping within itself the mark
of the past element" while at the same time being "vitiated by the mark of its
relation to the future element," thus "constituting what is called the present
by means of this very relation to what it is not."[40] Or as Luhmann puts it, "One
could say that meaning equips an actual experience or action with redundant
possibilities"—namely, what *was* selected (the actual) and what could have
been (the possible)—and this is crucial for any system's ability to respond to
environmental complexity by building up its own complexity via the form of
meaning; this is what Luhmann means when he says that *"this formal requirement refers meaning to the problem of complexity."*[41]

"The genesis and reproduction of meaning presupposes an infrastructure
in reality that constantly changes its states," Luhmann writes. "Meaning then
extracts differences (which only as differences have meaning) from this substructure to enable a difference-oriented processing of information. On all
meaning, therefore, are imposed a temporalized complexity and the com-
pulsion to a constant shifting of actuality, without meaning itself vibrating
in tune with that substructure."[42] From an adaptive and evolutionary point
of view, then, as Bruce Clarke examines in this volume, self-reference and
the form of meaning do not indicate solipsism. Quite the contrary. As Luhmann points out, it is "unproductive for meanings to circulate as mere self-
referentiality or in short-circuited tautologies. . . . One can think, 'This rose is a

rose is a rose is a rose.' But this use of a recursive path is productive only if it makes itself dependent on specific conditions and does not always ensue."[43] And herein lies the difference for Luhmann between *meaning* and *information*, one that recalls Derrida's emphasis in *Limited Inc.* on the specific pragmatics of iterability. Applying Spencer-Brown's "form of condensation," as explained in Schiltz's essay in this volume, Luhmann continues: "A piece of information that is repeated is no longer information. It retains its meaning in the repetition but loses its value as information. One reads in the paper that the deutsche mark has risen in value. If one reads this a second time in another paper, this activity no longer has value as information . . . although structurally it presents the same selection."[44] Something can be meaningful, in other words, but have no informational value.

One thus "begins not with identity but with difference"—with two differences, in fact: the difference inherent in every experience "between what is *actually given* and what can *possibly* result from it" that is given in the internal form of meaning itself; and the difference between *meaning* and *information* that is forced upon the system by environmental complexity and temporality. "Only thus can one give accidents informational value and thereby construct order, because information is nothing more than an event that brings about a connection between differences—'a difference that makes a difference.' Therefore, we encounter *the decomposition of meaning per se*," the "de-tautologization of meaning's self-reference" forced upon the system by the adaptive pressure of the environment, of the "outside world."[45] This is why—contrary to the view of systems theory as solipsistic, imperialistic, and so on—Luhmann insists that "The difference between meaning and world is formed for this process of the continual self-determination of meaning as the difference between order and perturbation, between information and noise. Both are, and both remain, necessary. The unity of the difference is and remains the basis for operation. *This cannot be emphasized strongly enough. A preference for meaning over world, for order over perturbation, for information over noise is only a preference.* It does not enable one to dispense with the contrary."[46]

In the form of meaning, then, we find that systems increase their contacts with their environments paradoxically by *virtualizing* them. "Meaning is the continual actualization of potentialities," Luhmann writes.

> But because meaning can be meaning only as the difference between what is actual at any given moment and a horizon of possibilities, every actualization always also leads to a virtualization of the potentialities that could be connected up with it. The instability of meaning resides in the untenability

of its core of actuality; the ability to restabilize is provided by the fact that everything actual has meaning only within a horizon of possibilities . . . [that] can and must be selected as the next actuality. . . . Thus one can treat the difference between actuality and possibility in terms of temporal displacement and thereby process indications of possibility with every (new) actuality. Meaning is the unity of actualization and virtualization, of re-actualization and re-virtualization, as a self-propelling process.[47]

This "virtualization" via meaning is an extraordinarily powerful evolutionary dynamic, and it is put to good use by both psychic and social systems. Indeed, as Luhmann insists, "Not all systems process complexity and self-reference in the form of meaning"—and here one could think of various biological systems[48]—"but for those that do, it is the *only* possibility. Meaning becomes for them the form of the world and consequently overlaps the difference between system and environment."[49] Or as Luhmann sometimes characterizes it—in a formulation whose resonances with Derrida's essays such as "Structure, Sign, and Play" are clear enough, I think—"the relationship between meaning and world can also be described with the concept of decentering. As meaning, the world is accessible everywhere: in every situation, in any detail," which is to say "that the world is indicated in all meaning. To that state of affairs corresponds an a-centric world concept," and hence "the closure of the self-referential order is synonymous here with the *infinite openness of the world*."[50]

This co-implication of psychic and social systems via the formal dynamics of meaning, combined with Luhmann's simultaneous insistence on the strict separation of psychic and social systems as discrete autopoietic entities, marks one of systems theory's most difficult and counterintuitive features—but also one of its most powerful innovations. In a formulation as matter-of-fact as it is beguiling, Luhmann writes: "Humans cannot communicate; not even their brains can communicate; not even their conscious minds can communicate. Only communication can communicate." "What we experience as our own mind operates as an isolated autopoietic system," Luhmann points out, and in fact, that isolation is "an indispensable condition of its possibility." There is "no conscious link between one mind and another," nor is there any "operational unity of more than one mind as a system"—all of which, Luhmann argues, is essentially taken for granted at this point by contemporary neurophysiology.[51] Indeed, he asks, how could any psychic system maintain its own functions if it shared its unity with other minds? How could I deliver a lecture if I shared the moment-to-moment ebb and flow of psychic activity of even one other consciousness in the room? In this sense, "communication," Luhmann writes, "operates with an unspecific reference to the participating state of mind; it is especially unspecific as to perception. It cannot copy states of mind, cannot imitate them, cannot represent them. This is the basis for the possibility of communication's building up a complexity of its own."[52]

Our intuitions, of course, would seem to suggest otherwise, and this is so precisely because psychic systems and social systems have co-evolved, each serving as the environment for the other, and this "has led to a common achievement, employed by psychic as well as social systems."[53] That achievement, of course, is meaning. "Meaning," Luhmann writes, "is the true 'substance' of this emergent evolutionary level. It is therefore false (or more gently, it is falsely chosen anthropocentrism) to assign the psychic . . . ontological priority over the social. It is impossible to find a 'supporting substance' for meaning. Meaning supports itself in that it enables its own self-referential reproduction. *And only the forms of this reproduction differentiate psychic and social structures*"—namely, "whether consciousness [in the case of psychic systems] or communication [social systems] is chosen as the form of operation."[54] Here, as I have already suggested, we find Luhmann's answer to Derrida's critique of the auto-affection of the voice and of consciousness as presence in *Speech and Phenomena, Of Grammatology,* and elsewhere: of the fallacy that writing or communication could be referred for its efficacy as a representation to an ontic substrate of consciousness and the psychic system, whereas in fact it is the ontologically unsupported ur-dynamic of writing (Derrida) or meaning (Luhmann) that is fundamental and that allows psychic and social systems to interpenetrate.

"The difficulty in seeing this," Luhmann writes (in a disarmingly common-sensical moment), "lies in that every consciousness that tries to do so is itself a self-referentially closed system and therefore cannot get outside of consciousness. For consciousness, even communication can only be conducted consciously and is invested in further possible consciousness. *But for communication this is not so.* Communication is only possible as an event that transcends the closure of consciousness: as the synthesis of more than the content of just one consciousness."[55] The confusing of consciousness and communication, if one wants to put it that way, is precisely why "the concept of meaning must be employed on such a high theoretical level. Meaning enables psychic and social systems to interpenetrate, while protecting their autopoiesis; meaning simultaneously enables consciousness to understand itself and continue to affect itself in communication, and enables communication to be referred back to the consciousness of the participants."[56]

The all-important medium that allows this "interpenetration" via the form of meaning to take place is, you will have already guessed, *language.* But "this does not mean language determines consciousness," Luhmann writes; "psychic processes are not linguistic processes," he continues, "nor is thought in any way 'internal dialogue' (as has been falsely maintained). It lacks an 'internal addressee.' There is no 'second I,' no 'self' in the conscious system, no 'me' vis-à-vis an 'I,' no additional authority that examines all linguistically formed thoughts to see whether it will accept or reject them and whose decision consciousness seeks to anticipate."[57] Luhmann's point here no doubt takes for granted similar formulations throughout Derrida's early work in *Speech and Phenomena*, *Of Grammatology*, and elsewhere, but the emphasis in Luhmann falls rather differently, on the evolutionary aspects of this disarticulation. What is important for Luhmann is that one must do justice to the powerful role of language in the co-evolution of psychic and social systems while simultaneously paying attention to their autopoiesis and self-referential closure. On the one hand, "the evolution of social communication is only possible in a constantly operative link with states of consciousness," which is provided by the medium of language;[58] on the other hand, language "transfers social complexity into psychic complexity" in a process generically referred to in contemporary theory as subjectivation.[59] "The social system places its own complexity, which has stood the test of communicative manageability, at the psychic system's disposal,"[60] but at the same time, language (and, even more, writing) ensures "for the communication system what Maturana calls the conservation of adaptation: the constant accommodation of communication to the mind. They define the free space of autopoiesis within the social communication system."[61]

For Luhmann, then, language is not constitutive of either psychic or social systems but is rather a very specific, second-order phenomenon—a type of "*symbolically generalized communication media*"[62]—that those systems use in the services of the *first*-order processes of meaning for maintaining their own autopoiesis while at the same time enabling them to interpenetrate and use each other's complexity to mutual benefit. From Luhmann's point of view, language is "not just a means of communication, because it functions in psychic systems without communication" in the strict sense of having to take place;[63] but at the same time, "communication is also possible without language" and may take place in all sorts of nonlinguistic ways, "perhaps through laughing, through questioning looks, through dress," and so on.[64]

In fact, what is fundamental about communication for Luhmann is not its (dis)relation to language, but rather that it is a "synthesis of three selections":[65] "information" (the "content," if you like, to be communicated); "utterance" (the specific, pragmatic communicative event or behavior selected to communicate information); and "understanding" (a receiver's processing of the difference between information and utterance that completes the communicative act).[66] Again, the issue is not just difference; *all* forms of meaning, of which communication is a specific instance, operate by means of difference; the issue is whether (to remember Gregory Bateson's phrase) an utterance is a "difference that makes a difference" in terms of the system's autopoiesis. Or as Luhmann puts it, "difference *as such* begins to work if and insofar as it can be treated as information in self-referential systems."[67] To recall Luhmann's earlier example of the value of the deutsche mark, an utterance, once repeated, may retain the same form as *meaning* but lose its status as *information*; it retains the same form but has lost its capacity to "select the system's states"—not because its form has changed but because the state of the system has.[68] This fact draws our attention, in turn, to what Derrida in *Limited Inc.* calls the "specific," "pragmatically determined" nature of any instance of undecidability, the emphasis upon which would seem to run counter to Luhmann's assertion that "communication is realized if and to the extent that understanding comes about."[69] Here again, however, Derrida and Luhmann converge upon the same point from opposite directions; while Derrida emphasizes the *final* undecidability of any signifying instance, Luhmann stresses that, even so, systems *must* decide; they must selectively process the difference between information and utterance if they are to achieve adaptive "resonance" with their environments. Thus, underneath this apparent divergence is a shared emphasis—against "relativism" and "anything goes" reflexivity—upon the determinate specificity of the signifying or communicative instance that must be negotiated, which is precisely why in *Limited Inc.* Derrida *rejects* the term "indeterminacy" because it occludes an understanding of the "*determinate* oscillation between possibilities (for example, of meaning, but also of acts)."[70]

Similarly, in Luhmann writing takes center stage as the paradigm of communication but only because it exemplifies a deeper "trace" structure (the grammé of the "programme," as it were) of meaning—a paradigm whose essential logic is only intensified by the sorts of later technical developments, beginning with printing, in which we have already seen Derrida himself keenly interested in texts like *Without Alibi* and *Archive Fever*. In this light, the problem with "oral speech," as Luhmann describes it, is that it threatens to collapse the difference between information and utterance, performatively subordinating the former to the latter and presuming their simultaneity—"leaving literally no time for doubt," as Luhmann puts it—in precisely the manner analyzed in Derrida's early critique of the subordination of writing to speaking.[71] But if the value of language is that it is "the medium that increases the understandability of communication beyond the sphere of perception,"[72] then writing is its full realization. "Only writing," Luhmann observes, "enforces the clear distinction between information and utterance," and "only writing and printing suggest communicative processes that react, not to the unity of, but to the difference between utterance and information.... Writing and printing enforce an experience of the difference that constitutes communication: they are, in this precise sense, more communicative forms of communication."[73]

Language, then, may be "a medium distinguished by the use of signs"—one that is capable of "*extending* the repertoire of understandable communication *almost indefinitely in practice*," an achievement whose significance "can hardly be overestimated." But "it rests, however, on functional specification. Therefore one must also keep its boundaries in view."[74] For Luhmann—and this is something like the negative image or reverse aspect of Derrida's early reading of Saussure, and specifically his drawing out of the full implications of Saussure's contention that language is a diacritical system that operates "without positive terms"—to subsume the dynamics of meaning under the theory of the sign is to ignore what he calls the "basal, recursive self-reference" that "forms the context in which all signs are determined."[75] Hence, "the concept of the symbolic generalization of meaning's self-reference replaces the concept of the sign that until now has dominated the theoretical tradition."[76]

For Luhmann, whether or not to understand Derrida precisely in terms of that theoretical tradition has been a matter of some uncertainty—an uncertainty that mirrors, to a large extent, broader disagreements in theory and philosophy about how Derrida is to be read and whether, moreover, the same understanding applies to his earlier versus later work.[77] At certain times, Luhmann suggests a high degree of translatability between the two theories, while at others he is concerned to keep his distance.[78] But my point here is not to rehearse these differences (much less to suggest which understanding of Derrida is "right"); nor is it to try and further systematize the relationship between Luhmann and Derrida along the lines already carried out quite ably by critics such as Dietrich Schwanitz, David Wellbery, Drucilla Cornell, Hans Ulrich Gumbrecht, and others. Rather, my point is to suggest that if systems theory needs deconstruction in the sense I touched upon at the outset, then deconstruction also needs systems theory to help carry out work toward which it has, in comparison, only gestured.

This complimentarity rests, as I have been arguing, upon two fundamental *disarticulations* in Luhmann that are at the core of Derrida's work as well: the disarticulation of psychic and social systems and, on an even more fundamental structural level, the disarticulation of the formal dynamics of meaning from language per se. In my view—and I am saying nothing here, I believe, with which

Derrida himself would disagree—it is on the basis of this double disarticulation that the ethical and political ambitions of deconstruction, and how they arise from a set of theoretical commitments, largely rest. Those ambitions are aptly expressed by Derrida at moments like the following one in the interview "Eating Well":

> If one reinscribes language in a network of possibilities that do not merely encompass it but mark it irreducibly from the inside, everything changes. I am thinking in particular of the mark in general, of the trace, of iterability, of *différance*. These possibilities or necessities, without which there would be no language, *are themselves not only human.* . . . And what I am proposing here should allow us to take into account scientific knowledge about the complexity of "animal languages," genetic coding, all forms of marking within which so-called human language, as original as it might be, does not allow us to "cut" once and for all where we would in general like to cut.[79]

At such moments, Derrida unfolds the implications of the point he first made, for U.S. audiences, in *Of Grammatology*: that the form (and force) of *différance*, the grammé, and the trace indicates a recursive, iterative dynamics of meaning that exceeds the rather tidy purview of human linguisticality alone; as Derrida puts it in *Of Grammatology*, "In all senses of the word, writing thus *comprehends* language."[80] And it is on the strength of that theoretical commitment that the ethical issues involved—in this particular case, issues related to what is popularly known as "animal rights"—arise.

Similarly, in his remarkable late essay on Lacan's rendering of the human/animal divide vis-à-vis the "subject of the signifier" and Lacan's contention that animals are incapable of "pretending to pretend," the *ethical* question of our obligations to nonhuman beings is generated by a *theoretical* articulation of the force of the trace (versus the Lacanian "signifier") that pushes Derrida's thought very much in the direction of Luhmann's work on the dynamics of meaning in autopoietic systems. As Derrida puts it there, "It is difficult to reserve, as Lacan does, the differentiality of signs for human language only, as opposed to animal coding. What he attributes to signs that, 'in a language' understood as belonging to the human order, 'take on their value from their relations to each other' and so on, and not just from the 'fixed correlation' between signs and reality, can and must be accorded to any code, animal or human." Moreover, "the structure of the trace," Derrida argues, "presupposes that *to trace* amounts to *erasing a trace* as much as to imprinting it. . . . How can it be denied that the simple substitution of one trace for another, the marking of their diacritical difference in the most elementary inscription—which capacity Lacan concedes to the animal—involves erasure as much as it involves the imprint?"—"*and this is why so long ago,*" Derrida adds, "*I substituted the concept of trace for that of signifier.*"[81]

Not only do such passages make it clear that Derrida is offering us not a theory of language, nor even of writing, but a far more ambitious, and thoroughly posthumanist, account of the paradoxical and deconstructive dynamics of meaning; they also make it clear that the account of meaning in systems theory should be viewed as the "reconstruction of deconstruction," one that provides the sort of rigorously articulated analysis toward which deconstruction only gestures philosophically, but for that very reason, in a sense, more provocatively than the "science" of Luhmann's sociology. This joining of forces between deconstruction and systems theory is crucial, I would like to think, not just from systems theory's vantage, but from deconstruction's as well. Derrida points toward this necessity in a very important footnote in *Positions*:

> The critique of historicism in all its forms seems to me indispensable. . . . The issue would be: can one criticize historicism in the name of something other than *truth and science* (the value of universality, omnitemporality, the infinity of value, etc.), and what happens to science when the *metaphysical* value of *truth* has been put into question, etc? How are the effects of science and of truth to be reinscribed? . . . Finally, it goes without saying that in no case is it a question of a *discourse against truth* or against science. (This is impossible and absurd, as is every heated accusation on this subject.) And when one analyzes systematically the value of truth . . . it is not in order to return naively to a relativist or sceptical empiricism.[82]

If we believe Gunther Teubner, such a perspective only draws into even sharper focus the need to supplement deconstruction with systems theory, whose explanatory force resides not only in a renovation of "science," which enables it to take account of self-reference and the manifold challenges of constructivism, but *also* in its ability to link these epistemological innovations to the historical emergence and specificity of particular social forms. Moreover, Teubner suggests, systems theory thus enables us to understand a crucial fact about social and political effectivity that in his view is lost on—or at least lost *in*—deconstruction: that the disclosure of paradox does not in itself threaten the autopoiesis of social systems, a point that in turn bears upon the putative political force of deconstruction's philosophical intervention. As Teubner puts it—and this would, I think, actually amount to taking seriously Derrida's insistence on the specific, pragmatically determined character of all instances of iteration and undecidability, now writ large—"Derrida's nightmare" is that "it is the secret of autopoiesis that social systems are no longer threatened by the paradoxes of their deconstructive reading. Autopoietic self-reproduction means that in routine operations they are constantly de-paradoxifying their foundational paradox. Thus, they are capable of deconstructing deconstruction, of course not in the sense that they can exclude it on a long-term basis but in the sense that they shift, displace, disseminate, historicize deconstruction itself, which drastically changes the conditions of its possibility."[83] And what this suggests for Teubner is that a deconstruction that took account of "the foundational paradoxes of emerging social systems would need to become historical, especially to recognize its own transformations. While the basic structures of the paradox remain the same, social processes of their invisibilization and the threatening moments of their re-emergence depend on historical contingencies. . . . The distinctions which are used for de-paradoxification," he continues, "are dependent on historical-societal conditions of plausibility, of acceptability, are contingent on binding knowledge in particular societies."[84]

Now one might well argue that Derrida's work—particularly his later investigations of questions of justice in relation to law, rights, and so on (both in his own work and in the work of his interlocutors)—is quite cognizant of this fact and indeed does what it does precisely to confront such systems of "binding knowledge" with paradoxes to which they must respond. But my larger point here is that the ahistorical, asociological character of deconstruction is not at all obviously a *failure* per se on Derrida's part, as Teubner would have it; indeed, it might well be viewed, from the vantage Derrida voices above on the "effects of truth," as a resolutely philosophical *refusal*. For Derrida's rejoinder to Teubner would no doubt be that systems theory—even on Luhmann's terms—cannot have its "science" and eat it too. The question, as Luhmann characterizes it, is that "an operation that uses distinctions in order to designate something we will call 'observation.' We are caught once again, therefore, in a circle: the distinction between operation and observation appears itself as an element of observation."[85] Empiricism, in other words, must always give way to contingent (and deconstructible) self-reference. From a Derridean point of view, then, the advantage that Teubner finds in Luhmann's historically oriented analysis would simply be referred back to an empiricism whose untenability Luhmann himself makes clear. Luhmann cannot maintain that "there exists no observer-independent, given reality"[86] and at the same time hold that "self-reference designates the unity that an element, a process, or a system is for itself. 'For itself' means independent of the cut of observation by others."[87] If it is indeed the case that "both attributions, observer attribution and object attribution, are possible," and that "the results can therefore be considered contingent,"[88] then this means, from a Derridean point of view, that the empiricism upon which any historicism depends is rendered permanently dysfunctional and that what we are really dealing with is a specific undecidability, in the domain of meaning, about what sorts of attributions are made, by whom and to whom, and with what particular effects. Thus, when Luhmann holds that "the difference between self-reference in the object and self-reference in the analysis, between the observed and the observing system, comes to be reflected in the problem of complexity,"[89] what this really means, in Derridean terms, is "comes to be reflected in the deconstructibility of the very distinctions upon which such a formulation depends."

Moreover, Derrida would surely be the first to argue that even if such distinctions are tenable in "analytical" terms (to take Luhmann's procedure at its word), when they come to be expressed *in language*, then our ability to draw clear boundaries between what Luhmann calls the "empirical," "analytical," and "semantic" dimensions of observation/description is only further eroded. More precisely, there are *three* orders of complexity here: the first order, which is the autopoietic self-reference of any system that makes self-reference and hetero-reference a product of its own self-referential closure; the second order, which includes autopoietic systems that, in addition to first-order self-reference, use the form of *meaning* to process complexity; and the third order, which includes autopoietic systems that, in addition to using basal self-reference and meaning, also use *language*. To acknowledge as much, from a Derridean point of view, is simply to take account of what we have already discussed as the "contaminating" force of iterability, which mitigates against the kind of conceptual ideality that would appear to be in play in Luhmann's assumption that the "empirical," "analytical," and "semantic" dimensions can be so neatly separated. Hence, Derrida insists in *Limited Inc.* that "there can be no rigorous analogy between a scientific theory . . . and a theory of language" and that, in fact, "it is more 'scientific' to take this limit, if it is one, into account and to treat it as a point of departure for rethinking this or that received concept of 'science' and of 'objectivity.'"[90]

What is involved here, then—to return to the text of Derrida's with which we began—is a certain difference between Derrida and Luhmann in relation to "the grammar of the event," a phrase that arises, as Peggy Kamuf points out in her introduction to *Without Alibi*, at the very moment of Derrida's entry into U.S. academic discourse. For Derrida insists, at that moment in 1966—as a matter of principle that appears to be maintained nearly forty years later—that "I don't know what a grammar of the event can be." As Kamuf surmises, Derrida responds "that he does not know what such a thing can be—except a reduction, a cancellation of the very thing being called 'event.'"[91] Of course, Luhmann would respond that the only way any of us are even around to declare such an inability *at all* is precisely on the basis of a prior "reduction" of environmental complexity, one that provides the autopoietic conditions of possibility for raising such questions (or *any* questions) in the first place. Or in Luhmann's words, "One must be capable of generating both continuity and discontinuity, which is easier in reality than in theory."[92]

Notes

1. In Haraway, *Simians, Cyborgs, and Women*, 43–68.
2. In this connection, my remarks here are a continuation of my attempt to align systems theory and post-structuralist theory more generally in my previous two books. In *Critical Environments*, this took the form not just of separate chapters devoted to Maturana and Varela and Luhmann (on the one hand) and Foucault and Deleuze (on the other), but also of an intensive analysis of Luhmann's and Deleuze's differences as they can be teased out by attention to the theoretical topography of "the fold." In my last book, *Animal Rites*, the focus was instead not primarily on Luhmann but on Maturana and Varela (and, to a lesser extent, Gregory Bateson) and how their evolutionary theory of the emergence of "linguistic domains" can help us to put some meat on the bones of Derrida's theory of the relationship between signification as a trace structure and the question of nonhuman and posthuman subjectivity.
3. Baecker, "Why Systems?" 61.
4. On this question of the (metaphysical) status of the voice, perhaps no triangulation is more instructive than that of Derrida, J. L. Austin, and Stanley Cavell, who is concerned to defend the voice against what he views as Derrida's own metaphysical temptations. For an overview of this triangulation and the stakes involved for the question of the subject in relation to different media, see my essay "When You Can't Believe Your Eyes."
5. Derrida, *Of Grammatology*, 9, 84.
6. Jacob, *The Logic of Life*, 1–2.
7. Derrida, *Of Grammatology*, 9.
8. Ibid., 324, n. 3.
9. Ibid., 84.
10. See the introduction to my edited collection, *Zoontologies*, xi.
11. Luhmann, "Deconstruction as Second-Order Observing," 101.
12. See Teubner, "Economics of Gift—Positivity of Justice."
13. Derrida, *Without Alibi*, 210, 72; emphasis in original. See also in this connection *Archive Fever*.
14. Derrida, *Without Alibi*, 72; emphases in original.
15. Ibid., 136; emphases in original.
16. Ibid., 135.
17. Ibid., 73; emphasis in original.

18. This is probably the point—on withdrawing the thinking of the event from an ontology—to insist once again on the difference between Derrida and Deleuze—or, for that matter and from a certain, more contemporary vantage, the difference between Luhmann and Varela. For an overview of these questions, see my *Critical Environments*, esp. 114–28; and, from another perspective, Hansen in this volume.

19. Luhmann, *Social Systems*, 67.

20. Ibid., 48.

21. Ibid., 215, 48.

22. Schwanitz, "Systems Theory according to Niklas Luhmann," 146.

23. Derrida, "Signature Event Context," 3.

24. Ibid., 7–8.

25. Derrida, "Différance," 64.

26. Derrida, "Speech and Phenomena," 26–27.

27. Derrida, "Afterword," in *Limited Inc.*, 119; emphases in original.

28. Ibid., 116, 148; emphases in original.

29. See Luhmann, *Social Systems*, 139, 145. Luhmann goes on to insist on his difference with Derrida's critique of this set of problems, but only by way of a quite reductive reading of Derrida's approach. I will return to this point, and the stakes involved in it, below.

30. Schwanitz, "Systems Theory according to Niklas Luhmann,"153.

31. Ibid., 156.

32. See Luhmann, *Social Systems*, 10: "Systems must cope with the difference between identity and difference when they reproduce themselves as self-referential systems; in other words, reproduction is the management of this difference. This is not a primarily theoretical but a thoroughly practical problem, and it is relevant not only for meaning systems."

33. Ibid., 25, 26.

34. "System differentiation," Luhmann writes, "is nothing more than the repetition within systems of the difference between system and environment. Through it, the whole system uses itself as environment in forming its own subsystems and thereby achieves greater improbability on the level of those subsystems by more rigorously filtering an ultimately uncontrollable environment. Accordingly, a differentiated system is no longer simply composed of a certain number of parts and the relations among them; rather, it is composed of a relatively large number of operationally employable system/environment differences, which each, along different cutting lines, reconstruct the whole system as the unity of subsystem and environment" (*Social Systems*, 7).

35. Think here, for example, of the Napster controversy and how changes in technology have forced a renegotiation of the "interpenetration" of economic and legal systems on the specific subsystemic site of intellectual property law—a classic example of how systems, following Luhmann's analysis of "interpenetration," make use of each other's own complexity to enhance their own for the purposes, if you like, of controlling (or at least "steering") the other.

36. Luhmann, *Social Systems*, 37, 33; emphasis in original.

37. As Teubner, a distinguished legal scholar, has pointed out, autopoietic social systems thrive on paradox in the sense that "de-paradoxification means to invent new distinctions which do not deny the paradox but displace it temporarily, and thus relieve it of its paralyzing power," so that, for example, "in European legal history, institutionalized distinctions between natural and positive law or, currently, distinctions between legislation and adjudication, have produced their impressive cultural achievements despite or precisely because of the legal paradox" ("Economics of Gift—Positivity of Justice," 32).

38. Luhmann, "The Paradox of Observing Systems," 83–84.

39. Luhmann, *Social Systems*, 60.

40. Derrida, "Différance," 65–66.

41. Luhmann, *Social Systems*, 60; emphasis in original.

42. Ibid., 63.

43. Ibid., 61.

44. Ibid., 67. Luhmann here cites Gregory Bateson, *Steps to an Ecology of Mind*.

45. Ibid., 75; emphases in original.

46. Ibid., 83; my emphasis.

47. Ibid., 65.

48. On this point, see Luhmann, *Social Systems*, 37: "On this basis one can then distinguish between, on the one hand, organic and neurophysiological systems . . . and, on the other, psychic and social systems, which are constituted by the production and processing of meaning." Luhmann's point is that both types of systems are self-referential, but for the latter, "meaning enables an ongoing reference to the system itself and to a more or less elaborated environment." In a now obsolete vocabulary, we would say that meaning enables a "representation" of the system/environment relation to the system itself (which is kept from *being* representationalist precisely by the inescapable fact of self-reference of all systems).

49. Ibid., 61; emphasis in original.

50. Ibid., 70, 62; emphasis in original.

51. Luhmann, "How Can the Mind Participate in Communication?" 169, 170.

52. Ibid., 178.

53. Luhmann, *Social Systems*, 59.

54. Ibid., 98; emphasis in original.

55. Ibid., 99; emphasis in original.

56. Ibid., 219.

57. Ibid.

58. Luhmann, "How Can the Mind Participate in Communication?" 173.

59. Luhmann, *Social Systems*, 272.

60. Ibid.

61. Luhmann, "How Can the Mind Participate in Communication?" 173. Though I cannot explore this here in any detail, it is worth noting, as Luhmann argues, that "The relationship of the accommodation of communication to the mind and the

unavoidable internal dynamics and evolution of society is also evident in the fact that changes in the forms in which language becomes comprehensible to the mind, from simple sounds to pictorial scripts to phonetic scripts and finally to print, mark thresholds of societal evolution that, once crossed, trigger immense impulses of complexity in a very short time" (Ibid., 174).

62. Luhmann, *Social Systems*, 161; emphasis in original.

63. Ibid., 94.

64. Ibid., 150.

65. Ibid., 147.

66. Ibid., 140–42, 147, 151.

67. Ibid., 40.

68. Ibid.

69. Ibid., 147.

70. Derrida, *Limited Inc.*, 148; emphasis in original.

71. Luhmann, *Social Systems*, 162.

72. Ibid., 160.

73. Ibid., 162–63.

74. Ibid., 160; emphases in original.

75. Ibid., 71.

76. Ibid., 94.

77. Here, very schematically, one would find Rodolph Gasché's reading of Derrida at one end of the spectrum and Richard Rorty's at the other. For a useful overview of these different ways of reading Derrida, see Rorty's "Derrida" and "Two Meanings," in *Essays on Heidegger*.

78. For the former, see, for example, Luhmann, *Art as a Social System*, 98–100, 157–58; for the latter, see, for example, Luhmann, "Deconstruction as Second-Order Observing," and *Social Systems*, 145–47, which seems to endorse Derrida's reading of Husserl in *Speech and Phenomena*, only to assimilate Derrida's work to "a theory of signs (language theory, structuralism)."

79. Derrida, "'Eating Well,'" 116–17; emphases in original.

80. Derrida, *Of Grammatology*, 7.

81. Derrida, "And Say the Animal Responded?" 126, 137; my emphases.

82. Derrida, "Positions," 104–5, n. 32.

83. Teubner, "Economics of Gift—Positivity of Justice," 36.

84. Ibid., 37–38.

85. Luhmann, "The Cognitive Program of Constructivism," 134.

86. Luhmann, *Observations on Modernity*, 19.

87. Luhmann, *Social Systems*, 33.

88. Luhmann, *Observations on Modernity*, 48.

89. Luhmann, *Social Systems*, 57.

90. Derrida, *Limited Inc.*, 118.

91. Kamuf in Derrida, *Without Alibi*, 6.

92. Luhmann, *Art as a Social System*, 156.

Complex Visuality

The Radical Middleground

IRA LIVINGSTON

He walked into the shreds of flame. But they did not bite into his flesh, they caressed him and engulfed him without heat or combustion. With relief, with humiliation, with terror, he understood that he too was an appearance, dreamt by another.
—BORGES, "The Circular Ruins"

I'm happy that the editors have allowed my rambling meditation to end this volume but without asking that I try to summarize, to round out the volume, or to provide any kind of *closure*, which—as this essay will amply demonstrate—I am neither able nor inclined to do. What follows, then, is instead a kind of loose but looping thread, of the kind that sticks out from the back of a complex tapestry. It is always a question as to whether a loose thread like this can simply be plucked off or whether—as I would like to think—it marks the ongoing vulnerabilities of a fabric (here, neocybernetics) to catch on other things (here, visual culture studies and ideology critique) and to alter the pattern. Of course such vulnerabilities can also be its *abilities*, the way it stays alive, grows and changes, engages and is engaged by its environment.

So here I try (and almost by definition fail) not to conclude but *to start over*, or as Foucault described his method, "to begin and begin again, to attempt and to be mistaken, to go back and rework everything."[1] One might even say that this account works as a rough definition of scientific method, at least insofar as scientific "progress of knowledge" narratives demand that all theories be shown, in the fullness of time, to be radically inadequate if not fundamentally wrong. In this version, science is a particularly productive way of being wrong, continually wrong, and of falling/failing forward.

But how does the notion of remaking and even starting over—or in terms of knowledge, of sustaining "beginner's mind"—jibe with the famously *conservative* nature of systems, which are mostly stuck with what they inherit and can never truly begin again? At issue here is the tension *between* emergence and embodiment—the tension that generates both this volume and the systems that neocybernetics engages. For example, "humans, as you call them" (to use William Burroughs's phrase) cannot renegotiate the most fundamental terms of our own embodiment;[2] we cannot "go back" and reject the evolution that has rendered us unable to fly (as a two-year-old I know has recently been having difficulty accepting) or decide that, after much consideration, we'd rather not be carbon-based life forms after all. But then again, how does this apparently uncontroversial account jibe with the narrative that we might someday transfer some essence of who we are into a silicon-based system (that is, a computer of some kind) or the recognition that in terms of our "extended phenotype," we have always been posthuman cyborgs?

A dialectical answer, already suggested above, might begin with the recognition that this kind of contradiction and complexity are not only those that threaten the integrity of a system (whether from the outside or inside), but also the principles of a system's organization and growth. In particular, though, I want to reject a tendency to situate (in one real place or one conceptual/disciplinary place) *an environment* of roiling, Wild West, edge-of-chaos complexity that acts as a fecund though dangerous matrix for the emergence of new systems (the space of birth or continual adolescence, as it were), and in another place (even if only an asymptotically approachable space), *the systems themselves* (where, one might expect, a kind of fully achieved embodiment or maturity must give way continually to deathly senility). To say "I want to" reject this is also to acknowledge a failure, the intensity of the desire marking the extent to which I remain structured by the dichotomy as a constitutive contradiction. Certainly anyone and everyone involved in the project of neocybernetics must recognize as a first principle the paradoxical dependence of openness on closure and vice versa and along with it the ongoing tension between emergence and embodiment. Call it a paradox or a contradiction or a tension or a dialectic—or simply a question. The question is both what drives neocybernetics to be born *and* where it gives way to other knowledge formations that will succeed it. Where I have in mind to look for an answer is in something like a kind of continual midlife crisis in which one is nonetheless most alive (my flagrant attempt to make my own condition exemplary). This project takes place in the no-man's land between emergence and embodiment and continues to be guided by a perversity that would refuse the coding of closure with science and masculinity and of openness with a feminized humanism/humanities.

Questions

What can an exploration of the development of contemporary visual culture tell us about notions of emergence, complexity, and systematicity? Is visuality a system, or is it being systematized in postmodern culture? How is visual complexity (the complexity of visual images themselves) related to complex visuality (the complexity of the multiple networks in which visuality participates)? I find the resolution of these questions lies in coming to grips with their fundamental ambiguity, which I characterize in visual terms as a radical middleground. Along the way, I argue that we must continue to displace the metaphor of visuality that has informed the neocybernetic notion of the observer.

Visual complexity—or complex visuality—seems to be a signature of postmodern culture.[3] Visual complexity seems to be as rife all around me—in the real streets and people and stores and signs of the city in which I live—as it is in the imagined worlds of film, television, and dreams. These observations suggest further, interlocking sets of questions: How are complexity and representations of complexity related? Are representations of complexity necessarily complex themselves, or is it possible to have simple representations of complexity? This latter could be the case when a merely quantitative complexity comes to stand for and even to displace the qualitatively complex. Arguably this is what happens in the "precession of simulacra," the ever-expanding generation and saturation of commodities (including images) driven by the logic of capitalism. But the relationship between quantity and quality is itself complex. Visual complexity seems also to be *coordinated with* and to *participate in* complex visuality—that is, complexity considered entirely within the visual may be a kind of real echo, a kind of aesthetic subroutine, of the complexity of the networks into which visuality is wired. *Coordination* and *participation* tend to displace *representation* all together. Finally, to what extent is the complexity we think we see an emergent phenomenon in itself, and to what extent is an emergent episteme—a *way* of seeing—conditioning us to notice and privilege a complexity that was to some extent always already present?

Instances

To begin with what comes most easily to hand, visual complexity may well be associated with multiple and rapidly changing and moving images, such as the fast-cut music videos often regarded as icons of postmodernism. The best claim to complexity in these cases is probably the continual perceptual and cognitive juggling and constellating that the barrage of images requires of the viewer. It seems funny now that modernist cultural critics have more often cast such a viewer as *passive*, usually in contradistinction to the supposedly more active reader of literary texts. More aptly for postmodernity, cultural theorist Jonathan Beller has proposed that the work required of the viewer is actually the exemplary form of labor under what he calls the Cinematic Mode of Production, a mode in which attention—especially visual attention—confers value on objects.[4] I will return to this idea below.

Visual complexity may also be fractal, involving orchestrations of pattern across scale. Here again, we can start by referring the complexity to the ocular and neurological fact of the multiple refocusings required of the viewer of an image that cannot be held together in a single glance. To take a familiar example, the fractal paintings of Jackson Pollock seem to me the height of a particularly formalist/modernist complexity. Pollock's expressionism evolved to cover the picture plane uniformly with an almost perfectly self-similar, abstract field—that is, his mature paintings retain the same density of detail at all scales from the very small to the large. This has been confirmed by computer analysis of the fractal dimensions of Pollock's paintings.[5]

Even so, it would seem that this uniformity makes such work less complex than other images—including the typically postmodern—that may be just as fractally dense but are also more heterogeneous. Such complexity is both enacted and thematized nicely in the iconic postmodern film *Blade Runner*, whose director spoke of making the film as assembling a "700-layer cake."[6] In addition to a *mise-en-scene* that is often visually busy and variously heterogeneous, the film offers several crucial scenes in which the small is conspicuously linked to the large by zooming in and refocusing—on a tiny origami figure, the minute details of a snapshot, the microscopic serial number printed around the base of a single artificial hair. This fractal layering in scale (a kind of complexity in itself) is linked both to a recursive loop whereby the film implicitly reflects on itself as an artifact of visual technology (this reflexivity constituting another kind of complexity) and, maybe somewhat more surprisingly, to an ongoing meditation in the film about the ambiguous and multiply crossed boundary between the natural and the artificial, a problematic undecidability that also yields another kind of complexity. Curiously, an ambiguated nature/culture boundary is also precisely that from which Jackson Pollock wove his fractal paintings, by working the physical properties of his paints (for example, their tendency, when flung, to form hairlike skeins or globby trails or to crackle as they dry). For us moderns anyway, there may be something almost archetypally complex—conceptually and aesthetically—about this kind of nature-culture hybridity.

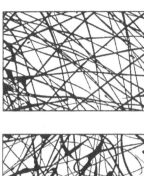

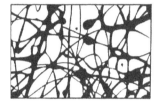

Detail (bottom) from Jackson Pollock's painting Number 14 (1948), compared with nonchaotic (top) and chaotic (middle) drip trajectories generated by the "Pollockiser" machine. (All sections c. 13 cm x 20 cm. By permission of the Pollock-Krasner Association/ Artists Rights Society, New York.)

Visual complexity may also involve what could be called visual hybridity or multi-modality. The most obvious examples here are interpenetrations of text and image and of reality and image. Here again the multiple modalities or dimensions of the image are linked to various perceptual and neurological modes that must be coordinated in order to engage them fully: imagine a pilot landing a plane, looking back and forth between multiple read-out screens and the real scene out the front window. Hybridity here might also refer to cases where more sophisticated kinds of visual "code-switching" are required, such as when multiple cultural and cross-cultural frameworks are involved; complexity here derives from the fact that we do not inhabit a single world with many things in it but multiple worlds, and we ourselves are necessarily multiple. (These points could again be illustrated by *Blade Runner*, famous for its visual mixing of a city that is both Los Angeles and New York, Asia and the West.)

But even in the relatively simple case of interpenetrations of reality and image, note that—in the case of films, for example—one would more accurately have to speak of the interpenetration of *images of reality* and *images of images*. This already suggests some of how visuality is itself complex: images add a recursive loop to the world they inhabit; even at the neurological level, this means that watching a film, for example, means coordinating being both awake and asleep, at least insofar as one's locomotor responses to filmed images are switched off, as in dreams. It is not so much the images themselves that are complex but that images make the world more complex, or in Niklas Luhmann's more general formula, "parts of the world (or for that matter any unity) have higher reflective potential than the unity itself."[7] It should be added here that visual imagery (and with it, the notion of consciousness-as-representation) can be considered a small addition to a world already built of multiple recursive loops, a world that never was a unity. In any case, though, it has become commonplace to refer to the ongoing explosion of images and image-making technologies as definitive for a modernity characterized by Heidegger as "the Age of the World Picture" and for a postmodernity construed by Debord's "Society of the Spectacle," Baudrillard's "Precession of Simulacra," or Beller's "Cinematic Mode of Production."

Throughout modernity and postmodernity, visuality has been increasingly separated out from other realms, increasingly commodified and mediated by assorted technologies. To take an early example, panel painting in the Renaissance helped to commodify art and contributed to prying it (literally and otherwise) away from the realms of religion and ritual; thus by the eighteenth century "aesthetics" was nameable as a separate realm unto itself. Visuality has been both increasingly distinguished as a realm unto itself and increasingly wired into the circuits of capital and social relations: modern systematization is marked by this simultaneous increase in the independence and interdependence of subsystems.

Definitions

Following the line of systems and autopoiesis theory, William Rasch cites biologist Robert Rosen's mandate "to view complexity not as an 'intrinsic property of a system [or] of a system description' but rather as an observer/observed relationship involving the choices that an observer makes, including the choice of what constitutes a system to begin with."[8] Accordingly, when I have wanted to establish the complexity of an image or an object (above), I have begun by coordinating it with complexity in a subject, starting with an ocular and neurological level.

Rasch adds Luhmann's account—of the evanescence of communication in a social system—that a system's "basic elements are not stable units (like cells or atoms or individuals) but events that vanish as soon as they appear."[9] The statement encapsulates one of post-structuralism's most transformative principles. Its general account of systematicity also conforms so perfectly to a basic description of visual media like film—that is, something whose elements are "events that vanish as soon as they appear"—that we are thrown back to historicizing again: the account seems to have been shaped by the cinematic age in which it was generated. Note further that Luhmann's statement can be taken as an ontological assertion of a general truth, something like the claim that "systems really exist in the world, as follows." But the statement can just as easily be taken as a definitional one, something more like "the term *system* shall be defined as follows." On one hand, then, it's a referential statement about the world; on the other, a self-referential statement about language. No doubt this apparent two-facedness is characteristic of both the statement *and the systems it describes*, and the exploration of this apparent resonance between observer and observed—in broader terms, the way language "bears *withness*" to the world—is one of the most paradigm-changing possibilities for the project of neocybernetics.[10] In any case, the notion of *performativity* offers a third way of understanding Luhmann's statement, one that displaces both reference and self-reference. What kinds of language games is the statement playing? What does it *do*, and with and against what other kinds of statements does it operate? For example, we might consider here how Luhmann's statement makes temporality itself—"eventness"—come to constitute complexity in relation to a kind of spatiality posited as a stable and knowable structure. This positing of *structure* as simple is a power move specific to how post-structuralism (the word itself no less than the theoretical movement it names) works to push structuralism into the past—precisely, one might say, to *historicize* it.

For Luhmann (going back to Rasch's account), "complexity, in turn, is seen as an observer's inability to define completely all these elements' connections and interactions. . . . Within the matrix of *observation as selection* there is no totalizing perspective or omniscient selector. Each act of observation is embedded in what it observes."[11] Again, this is a crucial principle, but thinking performatively about the statement—here, thinking what context it posits itself against—yields a more complex and contradictory picture. It is only against what I would call the *fantasy* of a disinterested and transcendently objective perspective—a fantasy that should be historically identified as specific to and definitive for imperialist modernity—that complexity can come to be defined as embeddedness, or rather, as the contradiction between transcendence and embeddedness. This contradiction has been characteristic of modern Western subjectivity at least since the late eighteenth century.[12]

Physicist and cellular automata guru Stephen Wolfram's definition of visual complexity also coordinates an observed and an observer in the act of description:

> In everyday language, when we say that something seems complex what we typically mean is that we have not managed to find any simple description of it—or at least of those features in which we happen to be interested. . . .
>
> When we are presented with a complex image, our eyes tend to dwell on it, presumably in an effort to give our brains a chance to extract a simple description.
>
> If we can find no simple features whatsoever—as in the case of simple randomness—then we tend to lose interest. But somehow the images that draw us in the most—and typically that we find most aesthetically pleasing—are those for which some features are simple for us to describe, but others have no short description that can be found by any of our standard processes of visual perception.[13]

Try to set aside, for the moment, the blithe trans-historical and transcultural universalizing of what "we" find pleasing. More interesting is what seems to be circular about this account—complexity is what interests us and what interests us is complex—and paradoxical: complexity is the necessary entanglement of the simple and the complex. This circularity and paradoxicality are not necessarily logical flaws but *traces* of the reflexive systems involved. The paradox can be developed by observing that if what we are interested in is complexity itself, then an image that we can easily identify as complex is thereby *less* complex than one whose complexity we find it difficult or impossible to ascertain.

Wolfram's approximate definition of visual complexity as the ability to capture and hold the eye and brain (a colloquial term for this is *velcro*) accords well with Beller's Cinematic Mode of Production (CMP), since the CMP's prime directive is to catch and hold attention—to make the viewer work, much as the flower makes the bee work. By the way, bees making honey is a common early modern metaphor for aesthetic labor, and it is more than a metaphor, really, since in both cases we are talking about a kind of co-evolution and interpenetration of systems. The flower becomes part of the bee's metabolism as the bee becomes part of the flower's reproductive system. Visuality becomes complex as more systems are wired through it and as it is wired through more other systems; as it becomes the switchboard for psyches, identities, economies; as culture and society increasingly reproduce visually and have a visual metabolism. For example, I have been repeatedly struck by how much this observation seems to apply to the U.S. invasion and occupation of Iraq (during which this essay was written and revised). Early on, it seemed very much like the image of the just-captured and haggard Saddam Hussein on the cover of American newspapers was the political capital for which billions of dollars and thousands of lives had been spent; later, the photographs from Abu Ghraib were both records and instruments of torture, images that "backfired" to undercut the U.S. representation of itself as a "liberator" and to fuel resistance to the occupation.

Middleground

If visual complexity has to be defined in part by its ability to interest us, we would do well to come to terms with Isabelle Stengers's observation that "'interest' actually derives from *interesse*, 'to be situated between.'" "To interest someone in something," Stengers contends, "means, first and above all to act in such a way that this thing—apparatus, argument, or hypothesis in the case of scientists—can concern the person, intervene in his or her life, and eventually transform it."[14] In the case of science, Stengers continues, interest works not just to configure scientists but to constellate "the multiple relations between scientific, social, industrial, and other interests," and thus "every interesting proposition redistributes the relations of signification, creates meaning but destroys it as well."[15] Notice that like the word *interest*, the word *meaning*—and the word *mediating*—also refer to a "being-in-the-middle" that brings us back to the notion of complexity as embeddedness. I'm tempted to say that at some level there is what might be called an anthropic principle at work here in the sense that we find ourselves—and life itself—necessarily in the middle of the universe, in terms of time, in terms of scale, and especially in terms of a temperature gradient, where the middle is an edge-of-chaos between order and disorder. Much more narrowly, though, I want to literalize Stengers's point and define visual complexity as a *middle-grounding*, as that which throws figure-and-ground into question. Note that by *ground* here I mean not simply that which is less interesting (in other words, that which is implied in the term *background*) but that against which a figure is made to appear, a frame or framework.

In recent tests of the visual perception of Asians and Westerners, members of both groups were shown simple computer-animated images of assorted fish swimming in a tank and asked to describe what they saw.[16] Westerners tended to identify a particular fish (the biggest one, generally) as more prominent and to make it the clear protagonist in their accounts, while Asians tended to describe collective relationships and patterns among the fish and their environment. To make this reductive characterization even more so, Westerners tended to stress figure, Asians ground.

Let's take this result at face value for the moment and ask what would tend to make the image more complex for *both* groups. Presumably this would be whatever interfered with the habitual foci of both groups in a way that (to use Wolfram's terms) made the usual short description impossible. The recipe for visual complexity, then, would involve whatever shaking, stirring, and cooking will make for an irreducibly heterogeneous middleground. If you aren't already using this principle when you make images, I invite you to try it. I want to call this a *radical* middleground to distinguish it from the kind of middleground that could easily be interpreted any way you like, since it compromises the figure/ground relationship in a way that ambiguates both. This is also the way I would like to approach neocybernetics' system/environment distinction.

If the complexity of our images is of a piece with the complexity of our world, this is not to say that our images accurately represent worldly complexity but that they participate directly in it in a way that displaces questions of representation all together, as subroutines wired into a networked and a *virtual* world, as Katherine Hayles has defined it: a world in which materiality is suffused with information. Historicization is paradoxical here in that we come to see that the world has always been virtual and are driven to ask why we didn't see it before, and thus to the narrative of a recovery *from* modernity, from the modernist fetish for purity and simple formalisms that keep trying to situate their figures of intelligibility against a ground of chaos. Bruno Latour's account is relevant here: we can recognize that we have never been modern insofar as the various nature/culture hybrids we have created have overrun our capacity to disavow them, a process (I would add) rather uncannily similar to the implosions of diversity back to the imperialist metropole. But we must also be suspicious of stories of evolution as increasing complexity (and following this, of modernity and postmodernity as increasingly complex systematizations and linkages at all scales) as artifacts of an ongoing reductionism that demands simple first principles and thus necessarily creates narratives of increasing complexity.

Neocybernetics

The attempt to use neocybernetic principles in approaching visuality (a project merely *posited* here) is complicated by the fact that the visual metaphor of *observation* has often been used in neocybernetics, and as is often the case, its metaphoricity tends to slip out of view. This section can be taken as part of the ongoing move to displace this metaphor, a move variously engaged by essays in this volume (I am particularly fond of Thompson's use of "sense-making"). Even so, the visual metaphor seems to me part of the ideological baggage that the project of neocybernetics must continue to unpack—or to carry.

Luhmann is especially clear, up front, on the metaphoricity of the notion of observation: "On the level of general systems theory, observation means nothing more than handling distinctions" and "self-observation is the introduction of the system/environment distinction within the system."[17] Accordingly (for example), single living cells—while they make no "observations" as such—do *handle distinctions*; in fact, they might well be fully describable as distinction-handling entities. Even so (for example, in Luhmann's sustained discussions of observation), the metaphor often seems to lose its implicit quotation marks and to become unmarked. What ideological or conceptual work does this slippage do?

Visual metaphors tend to bring the epistemological baggage of representationalism. This may happen in neocybernetics at some level even though, from its earliest usages in autopoiesis theory, the visual has also been used pointedly *against* representationalism and the naive notions of realism it can be made to underwrite. In the 1960s, Maturana participated in studies of frog vision that moved—from correlating retinal activity with external stimuli—to "understanding the nervous system as a closed network of interacting neurons" selectively *triggered by* but not *correlated with* an environment.[18] Notice that this paradigm shift develops structuralism's inaugural revelation that language be treated not in its referentiality to an outside world but as a system unto itself. As before, thinking in terms of *performativity* may be the simplest antidote to *both* representationalism and anti-representationalism.

Visual metaphors are also notorious for producing a false sense of distance or removal from what is being observed and, in turn, sneaking in a false sense of the observer's objectivity and even a transcendent perspective (that is, situating the observer as outside of the observed or, in other words, *desituating*). The sneakiness comes in because this action can be performed while being explicitly and even elaborately denied—as can be the case with universalizing statements such as "all perspectives are non-universal." Among the best antidotes continue to be *historicizing* (a sustained exploration of the historical embeddedness and contingency of all observations, including one's own) and *feminism* (which has understood visual privilege and the "god's eye view" as particular artifacts of patriarchy).

So far so good. But part of the impossibility of simply rejecting the representationalism and false sense of distance that tend to come with notions of

"observation" is that this rejection also tends to situate these kinds of epistemological mistakes as inherent in systems to begin with. Take, for example,
the claim that neocybernetics allows us to escape solipsism, meaning (presumably) the solipsism of a self that believes that it is a self and in its version of
the world as *the* world. The escape-from-solipsism account sounds good—
much in the same way that formulations of ethics based on recognition of
otherness sound good—and I would certainly want to affirm both principles
in themselves. But again, thinking performatively about such statements and
beginning to historicize them yields a contradictory and complex discursive
picture. Notice that such formulae leverage their knowledge claim (that is,
their claim to show us something counterintuitive) by *positing a world of naturally solipsistic creatures*. It is just along these lines that von Foerster gently
satirized his "American friends" of the early 1960s as excited by the revelation
that "I live in an Environment."[19] If, as von Foerster seems to be suggesting,
there is something particularly American about such a realization, laudable
in itself, it must be because imperially and patriarchally privileged American
subjectivity tends otherwise to assume what Keats once called, in reference to
William Wordsworth, the "egotistical sublime." As Clarke and Hansen point
out in the introduction to this volume (and then as Clarke goes on to elaborate),
after the revelation of an environment, at least two more self-reflexive steps
are necessary to inaugurate neocybernetics. First is the recognition that the
environment I posit is posited by me (or as Maturana and Varela put it, "anything said is said by an observer").[20] But can this really be the escape from
solipsism, the recognition that the world out there, and all of you people (and
especially you, dear reader), are figments of my imagination? On the contrary,
if this were all there was to it, this would be solipsism itself, and more than
just subjectivity (the posited "I") would remain to be displaced. The second
step is in coming to grips with the plurality—the always-already networked
nature—of self and mind: the social production of "stable yet mobile and
multiple recursive consensuses about shared environments," as Clarke defines von Foerster's "eigenvalues" (in this volume). Please notice that in the
move from the first to the second step, I've changed the metaphor from visual
("recognition") to tactile ("coming to grips"). This gesture is meant to signal
that to make the second step count, we must leave the realm of visual metaphors
of observation and representation, and with them too the realms of epistemology and of the self. What I want to do here is keep pushing Clarke's crucial
account of von Foerster—the account of solipsism not simply "transcended"
but revealed as a false problem in the first place, leaving us to start again if
we can.

The displacement of one self by many selves or one epistemology by another (or by many) always risks putting us (and by *us* I mean something like von Foerster's "American friends") back and back again into the paradox narrated by Isaac Bashevis Singer in his story of the snowfall in Chelm, the village of fools in Yiddish folktales. The elders of Chelm, believing that a sparkling treasure has fallen from the sky (recognizing an environment, one might say), meet to discuss how to take advantage of the gift. They agree that a messenger must be sent to warn people to stay in their houses lest they trample the treasure. But when they realize that the messenger will himself damage the treasure, they come to the happy solution of putting the messenger on a table and having him carried door-to-door by four men. When this plan is carried out and they see that the treasure has been trampled after all, they resolve that in the event of another snowfall, the four men carrying the table should themselves stand on tables carried by a total of sixteen *other* men. At this proposal— which is of course exactly what we hope the plurality of "second-order" cybernetics is *not*—the villagers rejoice in the wisdom of their elders.[21] Or shall we say the story is a warning to knowledge workers in imperialist nations to watch what our political feet are doing while we issue our epistemological warnings—or at least to be mindful of what continues to be "carried" by our metaphors?

If plurality or "promiscuous realism" (in John Dupré's formula) may only exacerbate the problem, maybe we can "start again" from a different place.[22] The notion of escaping solipsism, like the notion of an ethics based on recognition of otherness, is one of those things that make me feel I must be coming from another planet. Where I come from, I want to say, we started out as quite the opposite of solipsistic entities: we began as *figments of the dreams of others.* (This, by the way, is why I've used the Borges epigraph to begin this essay rather than situating it as a punchline.) In a world entirely defined and controlled by others, it was an epistemological leap to realize that we exist in some fashion, that we have needs and desires, that we can try to leverage a partial autonomy. And this leap continues to be all the more difficult and finally impossible since it is resisted at every turn, not only by others but also by ourselves insofar as we have been manufactured largely by the others and remain to a large extent their creatures. It is not that there are *simply* two worlds, one overprivileged one where people have the luxury of escaping their solipsism as a sort of ethical hobby, and another subaltern world in which people struggle to exist at all. In any case (just for starters), these are also contradictions that divide us from ourselves. Do you recognize any of yourself in these statements? The positioning of an "I" and a "we" and a "they" in these statements is just that, a performative positioning rather than a truth claim; all I really want to *do* with these statements is to open up the possibility of subaltern affiliation/recognition in neocybernetics.

To turn back for a moment to the introduction to this volume, Clarke and Hansen (following Luhmann and others) stake neocybernetics on several key distinctions: the system/environment distinction, the operational closure of systems, and the reduction of complexity that this closure enables. For example, they define neocybernetics in opposition to Katherine Hayles's statement that places "flows across borders" in the subject position; according to Hayles, flows "create complex dynamics of intermediation."[23] In their rejoinder, Clarke and Hansen put "the injunction against such flows"—presumably an injunction issued by a system—in the subject position. There is an easy way of splitting this difference, and my impulse to do so is related to the way my argumentative strategies take their cue from the systems that are my object (or is it vice versa?), which I think are better described as difference-splitters than difference-makers. In any case, there is no question that a system like a living body also *creates* multiple flows between itself and its environment (even if as a result of some injunction *against* flows) whereas a nonsystem like a rock is much more inert. A system is like an eddy in a stream; one can say that the eddy is a kind of injunction against the lateral flow of the stream or a kind of amplification of it. If this is one of those constitutive paradoxes of systems, we should be careful to continue to honor it as such, rather than using the distinction more baldly to separate neocybernetics from its "others."

Take a more extended example from contemporary visual culture: the rise of what has come to be called "reality television." Here I am merely "neocyberneticizing" a few of the observations from the excellent anthology *Understanding Reality Television*, a kind of translation I would like to see practiced much more broadly in visual culture studies and in neocybernetics.[24] (In the latter case, in other words, my fantasy is that students of neocybernetics would take various classic texts—of theory, history, and so forth—and translate them into neocybernetic terms—*hint, hint*—for example, as Clarke has just done so compellingly with narrative theory in *Posthuman Metamorphosis*.) At one level, in any case, it seems obvious that the emergence of reality TV has been generated by the self-reflexive closure of what we might call "the television system." This is especially clear insofar as (1) the system has always produced novelty through a very self-reflexive and ongoing generic mixing (in the case of reality TV, a mixing of documentary, soap opera, game show, etc.), and insofar as (2) the television system has always been defined and driven by the way it handles the distinction between celebrity and ordinariness. This distinction never has been that which separates the inside of television from the outside, but the contradiction at its heart, the way it meditates on its own production of ordinariness. This is fine as far as it goes, and part of what these "observations" do is to produce and handle distinctions that mark out an emergent academic field of "television studies." But what does it do to this field—and to the notion of a television system—to observe (as I began to do above) that the U.S. invasion of Iraq is also reality television—and here I don't mean that it is "like" reality television but that it *is*; an engineered intervention organized on fictional premises for purposes that include generating images and ideological narratives and revenue, mesmerizing people, and so on. For those who like their system boundaries crisp, this makes the notion of reality TV too broad and unwieldy; for the rest of us, it points toward something essential for understanding both reality TV and the U.S. invasion of Iraq.

Should we say, then, with Hayles, that the flows across borders between television and the world create a complex dynamics of intermediation in a world that must be said to have always been more or less virtual? Or should we say with Clarke and Hansen that the television system sustains itself via an injunction to selectively process the world, to radically reduce its complexity, and that our academic disciplines must do some version of the same and that, like it or not, we must come to some kind of terms with this? It seems obvious that, for starters, the only responsible answer is *both*.

My final example is a more pointedly reflexive one: neocybernetics itself. Let's begin by granting that there is a network of authors and texts (such as represented in this volume)—along with events such as panels at conferences, and if you prefer something more ethereal, *a set of ideas*—that can be called *neocybernetics*. Let's say that it has boundaries (in other words, that some authors and texts and ideas belong to it and others do not, even if both belonging and nonbelonging are complex and contradictory) and that it can reflect upon itself. We could say that it can claim anything it likes as part of itself and exclude anything or even that this claiming/excluding *is what it is*. For example, let it claim a history for itself that begins with systems theory, and let it disavow any other intellectual debts if it likes—a mystification one might find annoying, but never mind, since we are concerned here with the performative value of the disavowal, not its truth status. Is neocybernetics an emergent system? Will it continue to grow, pulling more and more authors, texts, and events into its constellation, becoming a more or less discrete discipline or meta-discipline? Or will it stabilize as a small but intrepid band passing down its wisdom to a few disciples? Or yet again, will it be recognized as fundamentally erroneous and be discarded, its practitioners shamed and then forgotten? Or in contradistinction to all of these rather melodramatic scenarios, will its insights be variously adopted, adapted, displaced into other knowledge formations?

I for one am so sure that *only this last case will come to pass* that I'm staking my eternal reputation on this prediction, which also has the virtue that it can relatively easily be verified (or at any rate falsified) by future generations, right here on the spot: *dear reader, was I right*? Do you identify neocybernetics as your dogma or discipline (does the word *neocyberneticist* appear under your name on your business card?), or do you scorn the entire project, having read this volume merely as the quaint and curious relic of some short-lived epistemological bubble? Or are you, like me, not quite what you would call a neocyberneticist but quite interested in what insights you can produce with something that could be called neocybernetics? Call it irony or postmodern identity if you like, this condition of being problematically both an insider and an outsider. And please excuse my universalism, but such a position seems to me the prerequisite for any kind of meaning or knowledge whatsoever.

It's easy enough to say that all this just means that neocybernetics is not a system or subsystem but a mere structure and that the real system is to be found at another level (for example, academia). This argument is always available (and always worth entertaining), but as in the case of visuality, as I have attempted to demonstrate here, we can best advance our understanding by pursuing the question of whether it is a system or not and by *not* deciding—that is, by recognizing and *sustaining* the undecidability. There is some danger that describing what systems do in terms of their ongoing success at sustaining undecidability is too blithe, so this account should probably be balanced by being recast as ongoing failure, something along the lines of Lacan's principle that "meaning indicates the direction in which it fails," Robert Frost's dictum that "like a piece of ice on a hot stove the poem must ride on its own melting," or some amplification of Luhmann's account of how systems "depend on constant disintegration."[25] Notice how far this account of surfing on one's own dissolution—a nice encapsulation of how dissipative systems work—takes us from the notion of systems that "perpetuate themselves." There are political and epistemological stakes in rejecting the Self-Perpetuating Paradigm.

To make a long story short, perpetuating themselves and policing their boundaries sounds an awful lot like what patriarchal nations and empires, and the subjectivities and systems modeled in their image, seek to do—just as "managing flows" seems like a mantra of neoliberal economics. Neocybernetics is necessarily shaped by the world and the history out of which it emerges, but can we leverage some choice as to what politics and ideology it will be made to serve? To bend neocybernetics to other than currently dominant political ends is, by definition, an uphill battle, one that requires sustained attention.

Notes

This essay owes a lot to Nick Mirzoeff, my mentor in visual culture studies; to Marty Babits, for unconsciously communicating to me his notion of the middleground of unconscious communication; and to Graham Sansom-Chasin, for his inspiring refusal to accept that humans can't fly.

1. Foucault, *Use of Pleasure*, 7.
2. Burroughs, "Ah Pook the Destroyer."
3. Mirzoeff, *The Visual Culture Reader*, 4–10.
4. See Beller, *The Cinematic Mode of Production*, or, for a nutshell version, see his "Kino-I, Kino-World."
5. Taylor, "Fractal Expressionism," 129–42.
6. Cited in Sammon, *Future Noir*, 47.
7. Luhmann, "Deconstruction as Second-Order Observing," 101.
8. Rasch, *Niklas Luhmann's Modernity*, 38–39.
9. Luhmann, *Political Theory in the Welfare State*, 83.
10. Livingston, *Between Science and Literature*, 4.
11. Rasch, *Niklas Luhmann's Modernity*, 47; my emphasis.
12. See Livingston, *Arrow of Chaos*, 84–104.
13. Wolfram, *A New Kind of Science*, 557, 559.
14. Stengers, *Power and Invention*, 83–84.
15. Ibid., 84.
16. See Nisbett, *The Geography of Thought*, 89–92.
17. Luhmann, *Social Systems*, 36–37.
18. Maturana, xv; emphasis in original.
19. Von Foerster, "On Constructing a Reality," 211.
20. Maturana and Varela, 8.
21. Singer, "The Snow in Chelm."
22. Dupré, *The Disorder of Things*.
23. Hayles, *My Mother Was a Computer*, 280.
24. Holmes and Jermyn, *Understanding Reality Television*.
25. Lacan, *Feminine Sexuality*, 150; Frost, *Collected Poems, Prose, and Plays*, 778; Luhmann, *Social Systems*, 48.

Bibliography

Ashby, W. Ross. *An Introduction to Cybernetics.* New York: J. Wiley, 1956.

Atlan, Henri. "Hierarchical Self-Organization in Living Systems: Noise and Meaning." In Zeleny, *Autopoiesis*, 185–208.

Baecker, Dirk, ed. *Kalkül der Form.* Frankfurt am Main: Suhrkamp, 1993.

———. "Knowledge and Ignorance." In Müller and Müller, *An Unfinished Revolution?* 337–50.

———, ed. *Problems of Form.* Trans. Michael Irmscher with Leah Edwards. Stanford: Stanford University Press, 1999.

———. "Why Systems?" *Theory, Culture & Society* 18, no. 1 (2001): 59–74.

Barbaras, R. "The Movement of the Living as the Originary Foundation of Perceptual Intentionality." In Petitot, Varela, Pachoud, and Roy, *Naturalizing Phenomenology*, 525–38.

Bateson, Gregory. *Steps to an Ecology of Mind.* New York: Ballantine, 1972.

Becchio, Cristina, and Cesare Bertone. "Beyond Cartesian Subjectivism: Neural Correlates of Shared Intentionality." *Journal of Consciousness Studies* 12, no. 7 (2005): 20–30.

Beller, Jonathan. *The Cinematic Mode of Production: Attention Economy and the Society of the Spectacle.* Hanover, N.H.: Dartmouth College Press, 2006.

———. "Kino-I, Kino-World: Notes on the Cinematic Mode of Production." In Mirzoeff, *The Visual Culture Reader*, 60–85.

Berger, Johannes. "Autopoiesis: Wie 'Systemisch' ist die Theorie Sozialer Systeme?" In *Sinn, Kommunikation und Soziale Differenzierung: Beiträge zu Luhmanns Theorie Sozialer Systeme.* Ed. Hans Haferkamp and Michael Schmidt, 129–52. Frankfurt am Main: Suhrkamp, 1987.

Berliner, Paul F. *Thinking in Jazz: The Infinite Art of Improvisation.* Chicago: University of Chicago Press, 1994.

Bitbol, Michel, and Pier Luigi Luisi. "Autopoiesis with or without Cognition: Defining Life at Its Edge." *Journal of the Royal Society Interface* 1, no. 1 (2004): 99–107.

Boden, Margaret A. "Autopoiesis and Life." *Cognitive Science Quarterly* 1 (2000): 117–45.

Borges, Jorge Luis. "The Circular Ruins." In *Labyrinths.* Ed. Donald A. Yates and James E. Irby, 45–50. New York: New Directions, 1964.

Bourgine, Paul, and John Stewart. "Autopoiesis and Cognition." *Artificial Life* 10, no. 3 (summer 2004): 327–45.

Brand, Stewart. "For God's Sake, Margaret: Conversation with Gregory Bateson and Margaret Mead." *CoEvolution Quarterly* 10 (summer 1976): 32–44.

Brauns, Jörg, ed. *Form und Medium*. Weimar: VDG, 2002.

Brewin, Chris R. *Posttraumatic Stress Disorder: Malady or Myth?* New Haven: Yale University Press, 2003.

Brier, Søren. "The Construction of Information and Communication: A Cyberse-miotic Reentry into Heinz von Foerster's Metaphysical Construction of Second-Order Cybernetics." *Semiotica* 154, nos. 1/4 (2005): 355–99.

Brose, Hans-Georg. "An Introduction towards a Culture of Non-Simultaneity." *Time and Society* 3, no. 1 (2005): 5–26.

Burroughs, William S. "Ah Pook the Destroyer / Brion Gyson's All-Purpose Bedtime Story." In Burroughs, *Dead City Radio* (CD). Fontana Island, 1990. Track 4.

Caruth, Cathy. *Trauma: Explorations in Memory*. Baltimore: Johns Hopkins University Press, 1995.

Castoriadis, Cornelius. *The Castoriadis Reader*. Trans. and ed. David Ames Curtis. Oxford: Blackwell, 1997.

———. "Entretien Cornelius Castoriadis et Francisco Varela." In Castoriadis, *Post-scriptum sur l'insignifiance*, 97–120.

———. *The Imaginary Institution of Society*. Trans. Kathleen Blarney. Cambridge, Mass.: MIT Press, 1987.

———. "The Logic of Magmas and the Question of Autonomy." In Castoriadis, *The Castoriadis Reader*, 290–318.

———. *Post-scriptum sur l'insignifiance: Entretiens avec Daniel Mermet, suivi de dialogue*. Paris: Éditions de l'Aube, 2000.

———. "Power, Politics, Autonomy." In Castoriadis, *Philosophy, Politics, Autonomy*. Trans. David Ames Curtis, 143–74. New York: Oxford University Press, 1991.

———. "Radical Imagination and the Society Instituting Imaginary." In Castoriadis, *The Castoriadis Reader*, 319–37.

———. "The State of the Subject Today." In Castoriadis, *World in Fragments: Writings on Politics, Society, Psychoanalysis, and the Imagination*. Trans. and ed. David Ames Curtis, 137–71. Stanford: Stanford University Press, 1997.

Chalmers, David J. *The Conscious Mind: In Search of a Fundamental Theory*. New York: Oxford University Press, 1996.

———. "Moving Forward on the Problem of Consciousness." *Journal of Consciousness Studies* 4 (1997): 3–46.

Chomsky, Noam. *Aspects of the Theory of Syntax*. Cambridge, Mass.: MIT Press, 1965.

Christis, Jac. "Luhmann's Theory of Knowledge: Beyond Realism and Constructivism?" *Soziale Systeme* 7, no. 2 (2001): 328–49.

Clam, Jean. "Die Grundparadoxie des Rechts und ihre Ausfaltung—Ein Beitrag zu einer Analytik des Paradoxen." In *Die Rückgabe des zwölften Kamels—Niklas Luh-*

mann in der Diskussion über Gerechtigkeit. Ed. Gunther Teubner, 109–43. Stuttgart: Lucius and Lucius, 2000.

———. "System's Sole Constituent, the Operation: Clarifying a Central Concept of Luhmannian Theory." *Acta Sociologica* 43 (2000): 63–79.

Clark, Andy. *Being There: Putting Brain, Body and World Together Again.* Cambridge, Mass.: MIT Press, 1997.

———. *Natural Born Cyborgs: Minds, Technologies, and the Future of Human Intelligence.* New York: Oxford University Press, 2004.

Clarke, Bruce. *Energy Forms: Allegory and Science in the Era of Classical Thermodynamics.* Ann Arbor: University of Michigan Press, 2001.

———. *Posthuman Metamorphosis: Narrative and Systems.* New York: Fordham University Press, 2008.

Clarke, Bruce, and Linda D. Henderson, eds. *From Energy to Information: Representation in Science and Technology, Art, and Literature.* Stanford: Stanford University Press, 2002.

Csordas, Thomas. "Embodiment and Cultural Phenomenology." In Weiss and Haber, *Perspectives on Embodiment,* 143–62.

Cull, Paul, and William Frank. "Flaws of Form." *International Journal of General Systems* 5 (1979): 201–11.

Damasio, Antonio. *The Feeling of What Happens: Body and Emotion in the Making of Consciousness.* New York: Harcourt Brace, 1999.

———. *Looking for Spinoza: Joy, Sorrow, and the Feeling Brain.* New York: Harcourt Brace, 2003.

Davis, Kingsley. "The Myth of Functional Analysis as a Special Method in Sociology and Anthropology." *Sociological Review* 24 (1959): 757–72.

Dennett, Daniel C. *Consciousness Explained.* Boston: Little Brown, 1991.

Derrida, Jacques. "And Say the Animal Responded?" In Wolfe, *Zoontologies,* 121–46.

———. *Archive Fever: A Freudian Impression.* Trans. Eric Prenowitz. Chicago: University of Chicago Press, 1996.

———. "Différance." Trans. Alan Bass. In Kamuf, *A Derrida Reader,* 59–79.

———. "'Eating Well' or The Calculation of the Subject." Trans. Peter Connor and Avital Ronnell. In *Who Comes after the Subject?* Ed. Eduardo Cadava, Peter Connor, and Jean-Luc Nancy, 96–119. New York: Routledge, 1991.

———. *Limited Inc.* Ed. Gerald Graff. Trans. Samuel Weber et al. Evanston, Ill.: Northwestern University Press, 1988.

———. *Of Grammatology.* Trans. Gayatri Chakravorty Spivak. Baltimore: Johns Hopkins University Press, 1976.

———. "Positions: Interview with Jean-Louis Houdebine and Guy Scarpetta." In Derrida, *Positions.* Trans. Alan Bass, 37–96. Chicago: University of Chicago Press, 1981.

———. "Signature Event Context." In Derrida, *Limited Inc.*

———. "Speech and Phenomena." Trans. David B. Allison. In Kamuf, *A Derrida Reader,* 6–30.

———. *Without Alibi.* Trans., ed., and intro. Peggy Kamuf. Stanford: Stanford University Press, 2002.

Dreyfus, Hubert L., and Stuart E. Dreyfus. "The Challenge of Merleau-Ponty's Phenomenology of Embodiment for Cognitive Science." In Weiss and Haber, *Perspectives on Embodiment,* 103–20.

Dupré, John. *The Disorder of Things: Metaphysical Foundations of the Disunity of Science.* Cambridge, Mass.: Harvard University Press, 1996.

Dupuy, Jean-Pierre. *Aux sources des sciences cognitives.* Paris: Éditions La Découverte, 1994.

———. *The Mechanization of the Mind: On the Origins of Cognitive Science.* Trans. M. B. DeBevoise. Princeton: Princeton University Press, 2000.

Dupuy, Jean-Pierre, and Francisco J. Varela. "Understanding Origins: An Introduction." In *Understanding Origins.* Ed. Francisco Varela and Jean-Pierre Dupuy, 1–25. Dordrecht: Kluwer, 1992.

Engstrom, Jack. "Precursors to *Laws of Form* in C. S. Peirce's Collected Papers." *Cybernetics and Human Knowing* 8, nos. 1–2 (2001): 25–66.

Esposito, Elena. "Ein zweiwertiger nicht-selbstständiger Kalkül." In Baecker, *Kalkül der Form,* 96–111.

Esterhammer, Angela. "The Cosmopolitan Improvvisatore: Spontaneity and Performance in Romantic Poetics." *European Romantic Review* 16 (2005): 153–65.

———. *Spontaneous Overflows and Revivifying Rays: Romanticism and the Discourse of Improvisation.* 2004 Garnett Sedgewick Memorial Lecture. Vancouver: Ronsdale Press, 2004.

Fernow, Carl Ludwig. "Über die Improvisatoren." In Carl Ludwig Fernow, *Römische Studien, Part 2,* 303–416. Zurich: Gessner, 1806.

Finscher, Ludwig, ed. *Die Musik in Geschichte und Gegenwart: Allgemeine Enzyklopädie der Musik begründet von Friedrich Blume Sachteil 4.* Kassel: Bärenreiter, 1996.

Fischer-Lichte, Erika. *Ästhetik des Performativen.* Frankfurt am Main: Suhrkamp, 2004.

Foucault, Michel. *The Use of Pleasure.* Trans. Robert Hurley. New York: Pantheon, 1985.

Fromm, Erich. *The Anatomy of Human Destructiveness.* New York: Holt, Rinehart, and Winston, 1973.

Frost, Robert. *Collected Poems, Prose, and Plays.* Ed. Richard Poirier and Mark Richardson. New York: Library of America, 1995.

Gadamer, Hans Georg. *Die Aktualität des Schönen: Kunst als Spiel, Symbol und Fest.* Stuttgart: Reclam, 1977.

Garvin, Paul L., ed. *Cognition: A Multiple View.* New York: Spartan Books, 1970.

Gioia, Ted. *The Imperfect Art: Reflections on Jazz and Modern Culture.* New York: Oxford University Press, 1988.

Godfrey-Smith, Peter. "Spencer and Dewey on Life and Mind." In *The Philosophy of Artificial Life.* Ed. Margaret A. Boden, 314–31. Oxford: Oxford University Press, 1996.

Grössung, Gerhard, Joseph Hartman, Werner Korn, and Albert Müller. *Heinz von Foerster 90*. Vienna: Echoraum, 2001.

Guattari, Félix. "Machinic Heterogenesis." In *Rethinking Technologies*. Ed. V. A. Conley, 13–27. Minneapolis: University of Minnesota Press, 1993.

Gumbrecht, Hans Ulrich. "Epiphany of Form: On the Beauty of Team Sports." *NLH* 30, no. 2 (spring 1999): 351–72.

———. "Form without Matter vs. Form as Event." *MLN* 111 (1996): 578–92.

———. *Production of Presence: What Meaning Cannot Convey*. Stanford: Stanford University Press, 2004.

Hamacher, Werner. "Das Beben der Darstellung." In *Positionen der Literaturwissenschaft: Acht Modellanalysen am Beispiel von Kleist's "Das Erdbeben in Chili."* Ed. David E. Wellbery, 149–73. Munich: C. H. Beck, 1993.

Hansen, Mark B. N. *Bodies in Code: Interfaces with Digital Media*. New York: Routledge, 2006.

Haraway, Donna J. *Simians, Cyborgs, and Women: The Reinvention of Nature*. New York: Routledge, 1991.

Hayles, Katherine. *How We Became Posthuman: Virtual Bodies in Cybernetics, Literature, and Informatics*. Chicago: University of Chicago Press, 1999.

———. *My Mother Was a Computer: Digital Subjects and Literary Texts*. Chicago: University of Chicago Press, 2005.

———. "Self-Reflexive Metaphors in Maxwell's Demon and Shannon's Choice: Finding the Passages." In Hayles, *Chaos Bound: Orderly Disorder in Contemporary Literature and Science*, 31–60. Ithaca, N.Y.: Cornell University Press, 1990.

Held, Klaus. "Husserl's Phenomenological Method." Trans. Lanei Rodemeyer. In *The New Husserl: A Critical Reader*. Ed. Donn Welton, 3–31. Bloomington and Indianapolis: Indiana University Press, 2003.

Henderson, Linda D. *The Fourth Dimension and Non-Euclidean Geometry in Modern Art*. Princeton: Princeton University Press, 1983.

Holmes, Emily A., Chris R. Brewin, and Richard G. Hennessy. "Trauma Films, Information Processing, and Intrusive Memory Development." *Journal of Experimental Psychology: General* 133, no. 1 (2004): 3–22.

Holmes, Su, and Deborah Jermyn, eds. *Understanding Reality Television*. London: Routledge, 2004.

Hurley, S. L. *Consciousness in Action*. Cambridge, Mass.: Harvard University Press, 1998.

Hutchins, Edwin. *Cognition in the Wild*. Cambridge, Mass.: MIT Press, 1994.

Jacob, Francois. *The Logic of Life: A History of Heredity*. Trans. Betty E. Spillmann. New York: Pantheon, 1973.

James, William. *The Principles of Psychology*. Cambridge, Mass.: Harvard University Press, 1981.

Jameson, Frederic. *Postmodernism, or, The Cultural Logic of Late Capitalism*. Durham, N.C.: Duke University Press, 1991.

Jonas, Hans. "Biological Foundations of Individuality." *International Philosophical Quarterly* 8 (1968): 231–51.

———. *Mortality and Morality: A Search for the Good after Auschwitz.* Evanston, Ill.: Northwestern University Press, 1996.

———. *The Phenomenon of Life: Toward a Philosophical Biology.* Chicago: University of Chicago Press, 1966.

Junge, Kay. "Medien als Selbstreferenzunterbrecher." In Baecker, *Kalkül der Form,* 112–51.

Kamuf, Peggy, ed. *A Derrida Reader: Between the Blinds.* New York: Columbia University Press, 1991.

Kant, Immanuel. *Critique of Judgement.* Trans. J. H. Bernard. New York: Hafner Press, 1951.

———. *Critique of Judgment.* Trans. Werner S. Pluhar. Indianapolis: Hackett Publishing, 1987.

Kauffman, Louis H. "The Mathematics of Charles Sanders Peirce." *Cybernetics and Human Knowing* 8, nos. 1–2 (2001): 79–110.

———. "On the Cybernetics of Fixed Points." *Cybernetics and Human Knowing* 8, nos. 1–2 (2001): 133–40.

———. "Self-Reference and Recursive Forms." *Journal of Social and Biological Structures* 10 (1987): 53–72.

Keller, Evelyn Fox. *A Feeling for the Organism: The Life and Work of Barbara McClintock.* New York: W. H. Freeman, 1984.

Kerouac, Jack. "On Spontaneous Prose." In *The Portable Jack Kerouac.* Ed. Ann Charters, 479–90. New York: Penguin Books, 1995.

Kim, Jaegwon. *Mind in a Physical World: An Essay on the Mind-Body Problem and Mental Causation.* Cambridge, Mass.: MIT Press, 1998.

Kirsch, Werner, and Dodo zu Knyphausen. "Unternehmen als 'autopoietische' Systeme?" In *Managementforschung I.* Ed. Wolfgang H. Staehle and Jörg Sydow, 75–101. Berlin: de Gruyter, 1991.

Kleist, Heinrich von. *An Abyss Deep Enough: Letters of Heinrich von Kleist with a Selection of Essays and Anecdotes.* Ed. and trans. Philip B. Miller. New York: E. P. Dutton, 1982.

Krasner, James. "Doubtful Arms and Phantom Limbs: Literary Portrayals of Embodied Grief." *PMLA* 119, no. 2 (2004): 218–32.

Kuhn, Thomas S. *The Structure of Scientific Revolutions.* Cambridge, Mass.: Harvard University Press, 1970.

Kurzweil, Ray. "Reflections on Stephen Wolfram's *A New Kind of Science.*" http://www.kurzweilai.net/articles/art0464.html.

Lacan, Jacques. *Feminine Sexuality: Jacques Lacan and the École Freudienne.* Ed. Juliet Mitchell and Jacqueline Rose. New York: Norton, 1982.

Landgraf, Edgar. "Self-Forming Selves: Autonomy and Artistic Creativity in Goethe and Moritz." *Goethe Yearbook* 11 (2002): 159–76.

Latour, Bruno. "Is *Re*-Modernization Occurring—And If So, How to Prove It?: A Commentary on Ulrich Beck." *Theory, Culture and Society* 20, no. 2 (2003): 35–48.

————. *We Have Never Been Modern.* Trans. Catherine Porter. Cambridge, Mass.: Harvard University Press, 1993.

LeDoux, Joseph. *The Emotional Brain: The Mysterious Underpinnings of Emotional Life.* New York: Simon and Schuster, 1996.

Leff, Harvey S., and Andrew F. Rex, eds. *Maxwell's Demon: Entropy, Information, Computing.* Princeton: Princeton University Press, 1990.

Lehmann, Maren. "Das Medium der Form—Versuch über die Möglichkeiten, George Spencer-Browns Kalkül der 'Gesetze der Form' als Medientheorie zu Lesen." In Brauns, *Form und Medium,* 39–56.

Leroi-Gourhan, André. *Gesture and Speech.* Trans. Anna Bostock. Cambridge, Mass.: MIT Press, 1993.

Letelier, J. C., G. Marín, and J. Mpodozis. "Autopoietic and (M,R) Systems." *Journal of Theoretical Biology* 222 (2003): 261–72.

Livingston, Ira. *Arrow of Chaos: Romanticism and Postmodernity.* Minneapolis: University of Minnesota Press, 1997.

————. *Between Science and Literature: An Introduction to Autopoetics.* Champaign: University of Illinois Press, 2006.

Llinás, Rodolpho. *I of the Vortex: From Neurons to Self.* Cambridge, Mass.: MIT Press, 2001.

Luhmann, Niklas. *Art as a Social System.* Trans. Eva M. Knodt. Stanford: Stanford University Press, 2000.

————. "The Autopoiesis of Social Systems." In *Sociocybernetic Paradoxes—Observation, Control and Evolution of Self-Steering Systems.* Ed. R. Felix Geyer and Johannes van der Zouwen, 172–92. London: Sage Publications, 1986.

————. "The Cognitive Program of Constructivism and a Reality That Remains Unknown." In Luhmann, *Theories of Distinction,* 128–52.

————. "The Control of Intransparency." *Systems Research in the Behavioral Sciences* 14 (1997): 359–71.

————. "Deconstruction as Second-Order Observing." In Luhmann, *Theories of Distinction,* 94–112.

————. *Die Gesellschaft der Gesellschaft.* Frankfurt am Main: Suhrkamp, 1997.

————. "Die Paradoxie der Form." In Baecker, *Kalkül der Form,* 197–212.

————. *Einführung in die Systemtheorie.* Ed. Dirk Baecker. Heidelberg: Carl-Auer-Systeme Verlag, 2002.

————. "Funktion und Kausalität." In Luhmann, *Soziologische Aufklärung 1—Aufsätze zur Theorie sozialer Systeme,* 9–30. Opladen: Westdeutscher Verlag, 1970 (1991).

————. "How Can the Mind Participate in Communication?" In Luhmann, *Theories of Distinction,* 169–84.

————. "The Medium of Art." In Luhmann, *Essays on Self-Reference,* 215–26. New York: Columbia University Press, 1990.

————. *Observations on Modernity.* Trans. William Whobrey. Stanford: Stanford University Press, 1998.

———. "The Paradox of Observing Systems." In Luhmann, *Theories of Distinction*, 73–93.

———. *Political Theory in the Welfare State*. Trans. John Bednarz, Jr. Berlin: Walter de Gruyter, 1990.

———. *Social Systems*. Trans. John Bednarz, Jr., with Dirk Baecker. Stanford: Stanford University Press, 1995.

———. "Temporalization of Complexity." In *Sociocybernetics*. Ed. R. Felix Geyer and Johannes van der Zouwen, vol. 2, 95–111. Leiden: Martinus Nijhoff, 1978.

———. *Theories of Distinction: Redescribing the Descriptions of Modernity*. Ed. William Rasch. Stanford: Stanford University Press, 2002.

———. "Weltkunst." In *Unbeobachtbare Welt: Über Kunst und Architektur*. Ed. Niklas Luhmann, F. D. Bunsen, and Dirk Baecker, 7–45. Bielefeld: C. Haux, 1990.

———. "What Is Communication?" In Luhmann, *Theories of Distinction*, 155–68.

Lutz, A., J.-P. Lachaux, J. Martinerie, and Francisco J. Varela. "Guiding the Study of Brain Dynamics by Using First-Person Data: Synchrony Patterns Correlate with Ongoing Conscious States during a Simple Visual Task." *Proceedings of the National Academy of Sciences* 99 (2002): 1586–91.

Margulis, Lynn. "Big Trouble in Biology: Physiological Autopoiesis versus Mechanistic Neo-Darwinism." In Margulis and Sagan, *Slanted Truths: Essays on Gaia, Symbiosis, and Evolution*, 265–82. New York: Springer-Verlag, 1997.

———. "The Conscious Cell." In *Cajal and Consciousness: Scientific Approaches to Consciousness on the Centennial of Ramon y Cajal's Textura*. Ed. P. C. Marijúan, 55–70. New York: New York Academy of Sciences, 2001.

———. *Symbiosis in Cell Evolution: Microbial Communities in the Archean and Proterozoic Eons*. 2d ed. New York: W. H. Freeman, 1992.

Margulis, Lynn, and Dorion Sagan. *Microcosmos: Four Billion Years of Evolution from Our Microbial Ancestors*. New York: Summit Books, 1986.

Maturana, Humberto R. "Autopoiesis." In Zeleny, *Autopoiesis*, 21–32.

———. "Biology of Cognition" (1970). In Maturana and Varela, *Autopoiesis and Cognition*, 2–58.

Maturana, Humberto, Jerome Lettvin, Warren McCulloch, and Walter Pitts. "Anatomy and Physiology of Vision in the Frog." *Journal of General Physiology* 43 (1960): 129–75.

Maturana, Humberto R., and Francisco J. Varela. *Autopoiesis and Cognition: The Realization of the Living*. Boston: Riedel, 1980.

———. "Autopoiesis: The Organization of the Living." In Maturana and Varela, *Autopoiesis and Cognition*, 59–141.

———. *Autopoietic Systems: A Characterization of the Living Organization*. Biological Computer Laboratory Report 9.4 (September 1, 1975).

———. *De Máquinas y seres vivos: Una teoría de la grganización biológica*. Santiago: Editorial Universitaria, 1994.

———. *The Tree of Knowledge: The Biological Roots of Human Understanding*. Rev. ed. Boston: Shambhala, 1998.

McCulloch, Warren S. *Embodiments of the Mind.* Cambridge, Mass.: MIT Press, 1975.

Meguire, Philip. "Discovering Boundary Algebra: A Simple Notation for Boolean Algebra and the Truth Functors." *International Journal of General Systems* 32, no. 1 (2003): 25–87.

Merleau-Ponty, Maurice. *Phénoménologie de la perception.* Paris: Gallimard, 1945.

———. *Phenomenology of Perception.* Trans. Colin Smith. London: Routledge, 1962.

Metzinger, Thomas. *Being No One: The Self-Model Theory of Subjectivity.* Cambridge, Mass.: MIT Press, 2003.

Mirzoeff, Nicholas, ed. *The Visual Culture Reader.* 2d ed. London: Routledge, 2002.

Moritz, Karl Philipp. *Werke.* Frankfurt am Main: Insel, 1981.

Morowitz, Howard. *The Emergence of Everything: How the World Became Complex.* Oxford: Oxford University Press, 2004.

Müller, Albert, and Karl H. Müller, eds. *An Unfinished Revolution? Heinz von Foerster and the Biological Computer Laboratory/BCL 1958–1976.* Vienna: Echoraum, 2007.

Nagel, Thomas. "What Is It Like to Be a Bat?" In Nagel, *Mortal Questions,* 165–80. New York: Cambridge University Press, 1979.

Nisbett, Richard E. *The Geography of Thought: How Asians and Westerners Think Differently . . . and Why.* New York: Free Press, 2003.

Noë, Alva. *Action in Perception.* Cambridge, Mass.: MIT Press, 2004.

Oyama, Susan. *The Ontogeny of Information: Developmental Systems and Evolution.* 2d ed. Durham, N.C.: Duke University Press, 2000.

Panksepp, Jaak. "The Periconscious Substrates of Consciousness: Affective States and the Evolutionary Origins of Self." *Journal of Consciousness Studies* 5 (1998): 566–82.

Patočka, Jan. *Body, Community, Language, World.* Trans. E. Kohák. Chicago and La Salle: Open Court, 1998.

Petitot, Jean, Francisco Varela, Bernard Pachoud, and Jean-Michel Roy, eds. *Naturalizing Phenomenology: Issues in Contemporary Phenomenology and Cognitive Science.* Stanford: Stanford University Press, 1999.

Piaget, Jean. *Biology and Knowledge: An Essay on the Relations between Organic Regulations and Cognitive Processes.* Trans. Beatrix Walsh. Chicago: University of Chicago Press, 1971.

Prigogine, Ilya, and Isabelle Stengers. *Order Out of Chaos: Man's New Dialogue with Nature.* New York: Bantam, 1984.

Quine, Willard von Orman. "On What There Is." In Quine, *From a Logical Point of View,* 1–19. Cambridge, Mass.: Harvard University Press, 1953.

Ramachandran, V. S. "Mirror Neurons and Imitation Learning as the Driving Force behind the 'Great Leap Forward' in Human Evolution." *Edge* 69 (June 2000). http://www.edge.org/3rd_culture/ramachandran/ramachandran_index.html.

Ramachandran, V. S., and Sandra Blakeslee. *Phantoms in the Brain.* New York: William Morrow, 1998.

Rasch, William. *Niklas Luhmann's Modernity: The Paradoxes of Differentiation.* Stanford: Stanford University Press, 2000.

Roberts, David. "Self-Reference in Literature." In Baecker, *Problems of Form*, 27–45.

Rorty, Richard. *Essays on Heidegger and Others: Philosophical Papers*, vol. 2. Cambridge: Cambridge University Press, 1991.

Rosch, Eleanor. "Principles of Categorization." In *Cognition and Categorization*. Ed. Eleanor Rosch and B. B. Lloyd. Hillsdale, N.J.: Lawrence Erlbaum, 1978.

Rosen, Robert. *Essays on Life Itself*. New York: Columbia University Press, 2000.

———. *Life Itself: A Comprehensive Inquiry into the Nature, Origin, and Fabrication of Life*. New York: Columbia University Press, 1991.

Roy, Jean-Michel, Jean Petitot, Bernard Pachoud, and Francisco Varela. "Beyond the Gap: An Introduction to Naturalizing Phenomenology." In Petitot, Varela, Pachoud, and Roy, *Naturalizing Phenomenology*, 1–80.

Rudrauf, David, Antoine Lutz, Diego Cosmelli, Jean-Philippe Lachaux, and Michel Le Van Quyen. "From Autopoiesis to Neurophenomenology: Francisco Varela's Exploration of the Biophysics of Being." *Biological Research* 36, no. 1 (2003): 27–65.

Sammon, Paul M. *Future Noir: The Making of Blade Runner*. London: Orion, 1996.

Schiltz, Michael. "Form and Medium—A Mathematical Reconstruction." In *Image and Narrative—Special Issue: Medium Theory*. Ed. Michael Boyden, Michael Schiltz, Jan Van Looy, and Gert Verschraegen, 2003. http://www.imageandnarrative.be/mediumtheory/michaelschiltz.htm.

Schiltz, Michael, and Makoto Nishibe. "Money between Form and Medium." *Evolutionary and Institutional Economics Review*. Forthcoming.

Schiltz, Michael, and Gert Verschraegen. "Spencer-Brown, Luhmann and Autology." *Cybernetics and Human Knowing* 9, nos. 3–4 (2002): 55–78.

Schneider, Eric D., and Dorion Sagan. *Into the Cool: Energy Flow, Thermodynamics, and Life*. Chicago: University of Chicago Press, 2005.

Schwanitz, Dietrich. "Systems Theory according to Niklas Luhmann—Its Environment and Conceptual Strategies." *Cultural Critique* 30 (spring 1995): 137–70.

Sheets-Johnstone, Maxine. *The Primacy of Movement*. Philadelphia: John Benjamins Press, 1999.

Silberstein, Michael, and John McGeever. "The Search for Ontological Emergence." *Philosophical Quarterly* 49, no. 195 (1999): 182–200.

Simon, Fritz B. "Mathematik und Erkenntnis: Eine Möglichkeit, die 'Laws of Form' zu lesen." In Baecker, *Kalkül der Form*, 38–57.

Simondon, Gilbert. *Du mode d'existence des objets techniques*. Paris: Aubier, 1989.

———. "The Genesis of the Individual." Trans. Mark Cohen and Sanford Kwinter. In *Incorporations*. Ed. Jonathan Crary and Sanford Kwinter, 296–319. New York: Zone, 1992.

———. *L'Individuation psychique et collective*. Paris: Aubier Montaigne, 1992.

———. *L'Individu et sa genèse physico-biologique*. Paris: PUF, 1964.

Singer, Isaac Bashevis. "The Snow in Chelm." In Singer, *Zlateh the Goat and Other Stories*, 29–38. New York: Harper Collins, 1966.

Smith, Hayel, and Roger Dean. *Improvisation, Hypermedia and the Arts since 1945*. Performing Arts Studies, vol. 4. Amsterdam: Harwood Academic Publishers, 1997.

Spencer-Brown, George. *Laws of Form.* London: Allen and Unwin, 1969; rp. Portland: Cognizer, 1994.

——. *Laws of Form.* New York: Dutton, 1979.

——. *Probability and Scientific Inference.* London: Longmans Green, 1957.

Stengers, Isabelle. *Power and Invention: Situating Science.* Trans. Paul Bains. Minneapolis: University of Minnesota Press, 1997.

Stewart, John. "Cognition = Life: Implications for Higher-Level Cognition." *Behavioural Processes* 35 (1996): 311–26.

——. "Life = Cognition: The Epistemological and Ontological Significance of Artificial Life." In *Toward a Practice of Autonomous Systems: Proceedings of the First European Conference on Artificial Life.* Ed. Paul Bourgine and Francisco J. Varela, 475–83. Cambridge, Mass.: MIT Press, 1992.

Stiegler, Bernard. *Technics and Time.* Vol. 1: *The Fault of Epimetheus.* Trans. Richard Beardsworth and George Collins. Stanford: Stanford University Press, 1998.

Taylor, Richard. "Fractal Expressionism—Where Art Meets Science." In *Art and Complexity.* Ed. John Casti and Anders Karlqvist, 117–44. Amsterdam: Elsevier, 2003.

Teubner, Gunther. "Economics of Gift—Positivity of Justice: The Mutual Paranoia of Jacques Derrida and Niklas Luhmann." *Theory, Culture and Society* 18, no. 1 (2001): 29–47.

——. *Recht als autopoietisches System.* Frankfurt am Main: Suhrkamp, 1989.

Teubner, Gunther, and Alberto Febbrajo, eds. *State, Law, and Economy as Autopoietic Systems: Regulation and Autonomy in a New Perspective.* Milan: Giuffre, 1987.

Thompson, Evan. *Mind in Life: Biology, Phenomenology, and the Sciences of Mind.* Cambridge, Mass.: Harvard University Press, 2007.

Thompson, Evan, and Francisco J. Varela. "Radical Embodiment: Neuronal Dynamics and Consciousness." *Trends in Cognitive Science* 5 (2001): 418–25.

Thrift, Nigel. "Remembering the Technological Unconscious by Foregrounding Knowledges of Position." *Environment and Planning D: Society and Space* 22 (2004): 175–90.

Timberg, Craig. "A Culture Vanishes in Kalahari Dust." *Washington Post,* June 3, 2005.

Turner, J. Scott. *The Extended Organism: The Physiology of Animal-Built Structures.* Cambridge, Mass.: Harvard University Press, 2000.

Varela, Francisco J. "Autonomy and Autopoiesis." In *Self-Organizing Systems: An Interdiciplinary Approach.* Ed. G. Roth and H. Schwegler, 14–24. New York: Campus Verlag, 1981.

——. "Autopoïese et émergence." In *La Complexité, vertiges et promesses.* Ed. Réda Benkirane. Paris: Le Pommier, 2002.

——. "A Calculus for Self-Reference." *International Journal of General Systems* 2 (1975): 5–24.

——. "Describing the Logic of the Living: The Adequacy and Limitations of the Idea of Autopoiesis." In Zeleny, *Autopoiesis,* 36–48.

——. "The Early Days of Autopoiesis: Heinz and Chile." *Systems Research* 13, no. 3 (1996): 407–16.

———. *Ethical Know-How: Action, Wisdom and Cognition.* Stanford: Stanford University Press, 1999.

———. "Intimate Distances: Fragments for a Phenomenology of Organ Transplantation." In *Between Ourselves: Second-Person Issues in the Study of Consciousness.* Ed. Evan Thompson, 259–71. Thorverton, UK: Imprint Academic, 2001.

———. "Introduction: The Ages of Heinz von Foerster." In von Foerster, *Observing Systems,* 2d ed., xiii–xviii. Salinas, Calif.: Intersystems, 1984.

———. "Making It Concrete: Before, during and after Breakdowns." In *Revisioning Philosophy.* Ed. James Ogilvy. Albany: SUNY Press, 1992.

———. "The Naturalization of Phenomenology as the Transcendence of Nature: Searching for Generative Mutual Constraints." *Alter* 5 (1997): 355–81.

———. "Neurophenomenology: A Methodological Remedy for the Hard Problem." *Journal of Consciousness Studies* 3 (1996): 330–50.

———. "Not One, Not Two." *CoEvolution Quarterly* 12 (fall 1976): 62–67.

———. "On Being Autonomous: The Lessons of Natural History for Systems Theory." In *Applied Systems Research.* Ed. George Klir, 77–85. New York: Plenum Press, 1977.

———. "On Observing Natural Systems (interview by Donna Johnson)." *CoEvolution Quarterly* 11 (summer 1976): 26–31.

———. "Organism: A Meshwork of Selfless Selves." In *Organism and the Origin of Self.* Ed. A. Tauber, 79–107. Dordrecht: Kluwer Academic Publishers, 1991.

———. "Patterns of Life: Intertwining Identity and Cognition." *Brain and Cognition* 34 (1997): 72–87.

———. *Principles of Biological Autonomy.* New York: Elsevier North Holland, 1979.

———. "Reflections on the Chilean Civil War." *Lindisfarne Letter* 8 (winter 1979): 13–19.

———. "Resonant Cell Assemblies: A New Approach to Cognitive Functions and Neuronal Synchrony." *Biological Research* 28 (1995): 81–95.

———. "The Specious Present: A Neurophenomenology of Time Consciousness." In Petitot, Varela, Pachoud, and Roy, *Naturalizing Phenomenology,* 266–314.

———. "Steps to a Science of Inter-Being: Unfolding the Dharma Implicit in Modern Cognitive Science." In *The Psychology of Awakening.* Ed. S. Bachelor, G. Claxton, and G. Watson, 71–89. New York: Random House, 1999.

———. "Whence Perceptual Meaning? A Cartography of Current Ideas." In Dupuy and Varela, *Understanding Origins,* 235–63.

Varela, Francisco J., and Antonio Coutinho. "Immunoknowledge: The Immune System as a Learning Process of Somatic Individuation." In *Doing Science: The Reality Club.* Ed. John Brockman, 237–56. New York: Prentice Hall, 1991.

Varela, Francisco J., and Natalie Depraz. "At the Source of Time: Valence and the Constitutional Dynamics of Affect." *Journal of Consciousness Studies* 12, nos. 8–10 (2005): 61–81.

Varela, Francisco J., F. J. Lachaux, J.-P. Rodriguez, and J. Martinerie. "The Brainweb: Phase Synchronization and Large-Scale Integration." *Nature Reviews: Neuroscience* 2 (2001): 229–39.

Varela, Francisco J., and Humberto M. Maturana. "Time Course of Excitation and Inhibition in the Vertebrate Retina." *Experimental Neurology* 26 (1970): 53–59.

Varela, Francisco J., Humberto M. Maturana, and Ricardo Uribe. "Autopoiesis: The Organization of Living Systems, Its Characterization and a Model." *BioSystems* 5 (1974): 187–96.

Varela, Francisco J., Evan Thompson, and Eleanor Rosch. *The Embodied Mind: Cognitive Science and Human Experience.* Cambridge, Mass.: MIT Press, 1991.

Varga von Kibéd, Matthias, and Rudolf Matzka. "Motive und Grundgedanke der 'Gesetze der Form.'" In Baecker, *Kalkül der Form,* 58–85.

Velvet Goldmine. Dir. Todd Haynes. Miramax, 1998.

Viskovatoff, Alex. "Foundations of Niklas Luhmann's Theory of Social Systems." *Philosophy of the Social Sciences* 29, no. 4 (1999): 481–516.

von Foerster, Heinz. "A Circuitry of Clues to Platonic Ideation." In *Aspects of the Theory of Artificial Intelligence.* Ed. Charles A. Muses, 43–82. New York: Plenum Press, 1962.

———. "Cybernetics of Epistemology." In von Foerster, *Understanding Understanding,* 237–41.

———. *Das Gedächtnis: Eine Quantum-physikalische Untersuchung.* Vienna: Franz Deuticke, 1948.

———. "Ethics and Second-Order Cybernetics." In von Foerster, *Understanding Understanding,* 287–304.

———. "For Niklas Luhmann: 'How Recursive is Communication?'" In von Foerster, *Understanding Understanding,* 305–23.

———. "Introduction to Natural Magic." In von Foerster, *Understanding Understanding,* 325–38.

———. "*Laws of Form.*" In *The Last Whole Earth Catalog.* Ed. Stewart Brand, 14. Palo Alto, Calif.: Portola Institute, 1971.

———. "Molecular Ethology, An Immodest Proposal for Semantic Clarification." In von Foerster, *Understanding Understanding,* 133–67.

———. "Notes on an Epistemology for Living Things." In von Foerster, *Understanding Understanding,* 247–59.

———. "Objects: Tokens for (Eigen-)Behaviors." In von Foerster, *Understanding Understanding,* 261–71.

———. *Observing Systems.* Salinas, Calif.: Intersystems Publications, 1981.

———. "On Constructing a Reality." In von Foerster, *Understanding Understanding,* 211–27.

———. "On Self-Organizing Systems and Their Environments." In von Foerster, *Understanding Understanding,* 1–19.

———. "Principles of Self-Organization in a Socio-Managerial Context." In *Self-Organization and Management of Social Systems.* Ed. H. Ulrich and Gilbert J. B. Probst, 2–24. Berlin: Springer, 1984.

———. *Understanding Understanding: Essays on Cybernetics and Cognition.* New York: Springer, 2003.

———. "What Is Memory That It May Have Hindsight and Foresight as Well?" In von Foerster, *Understanding Understanding*, 101–31.

von Foerster, Heinz, with Bernhard Poerksen. *Understanding Systems: Conversations on Epistemology and Ethics*. New York: Kluwer Academic/Plenum Publishers, 2002.

Wagner, Gerhard. "The End of Luhmann's Social Systems Theory." *Philosophy of the Social Sciences* 27 (1997): 387–409.

Weber, Andreas, and Francisco J. Varela. "Life after Kant: Natural Purposes and the Autopoietic Foundations of Biological Individuality." *Phenomenology and the Cognitive Sciences* 1, no. 2 (2002): 97–125.

Weiss, Gail, and Honi Fern Haber, eds. *Perspectives on Embodiment: The Intersection of Nature and Culture*. New York: Routledge, 1999.

Wellbery, David. "Contingency." In *Neverending Stories: Toward a Critical Narratology*. Ed. Ann Fehn, Ingeborg Hoesterey, and Maria Tatar, 237–57. Princeton: Princeton University Press, 1992.

Wheeler, Michael. "Cognition's Coming Home: The Reunion of Life and Mind." In *Proceedings of the Fourth European Conference on Artificial Life*. Ed. Phil Husbands and Inman Harvey, 10–19. Cambridge, Mass.: MIT Press, 1997.

Wiener, Norbert. *Cybernetics, or Control and Communication in the Animal and the Machine*. 2d ed. New York: MIT Press, 1961.

———. *The Human Use of Human Beings: Cybernetics and Society*. Boston: Houghton Mifflin, 1950.

Wittgenstein, Ludwig. *Logisch-philosophische Abhandlung: Tractatus Logico-philosophicus*. Ed. Brian McGuinness and Joachim Schulte. Frankfurt am Main: Suhrkamp, 1922 (1989).

———. *Tractatus logico-philosophicus—The German Text of Ludwig Wittgenstein's Logisch-philosophische Abhandlung*. London: Routledge and Kegan Paul, 1921 (1971).

Wolfe, Cary. *Animal Rites: American Culture, the Discourse of Species, and Posthumanist Theory*. Chicago: University of Chicago Press, 2003.

———. *Critical Environments: Postmodern Theory and the Pragmatics of the "Outside."* Minneapolis: University of Minnesota Press, 1998.

———. "When You Can't Believe Your Eyes: The Prosthetics of Subjectivity and the Ethical Force of the Feminine in *Dancer in the Dark*." *Subject Matters* 3, no. 2–4, no. 1 (2007): 113–44.

———, ed. *Zoontologies: The Question of the Animal*. Minneapolis: University of Minnesota Press, 2003.

Wolfram, Stephen. *A New Kind of Science*. Champaign/Urbana: Wolfram Media, 2002.

Wordsworth, William. *The Letters of William and Dorothy Wordsworth*. Vol. 1. Ed. Ernest de Selincourt. 2d ed., rev. Chester L. Shaver. Oxford: Clarendon Press, 1967.

Young, Iris Marion. *Throwing like a Girl and Other Essays in Feminist Philosophy and Social Theory*. Bloomington: Indiana University Press, 1990.

Yovits, Marshall C., and Scott Cameron, eds. *Self-Organizing Systems*. New York: Pergamon Press, 1960.

Zahavi, Dan. *Self-Awareness and Alterity: A Phenomenological Investigation*. Evanston, Ill.: Northwestern University Press, 1999.

Zeleny, Milan. "Autopoiesis: A Paradigm Lost?" In *Autopoiesis, Dissipative Structures, and Spontaneous Social Orders*. Ed. Milan Zeleny, 3–43. Boulder: Westview Press, 1980.

———, ed. *Autopoiesis: A Theory of Living Organization*. New York: North Holland, 1981.

Zolo, Danilo. "The Epistemological Status of the Theory of Autopoiesis and Its Applications to the Social Sciences." In Teubner and Febbrajo, *State, Law, and Economy*, 67–124.

———. "Function, Meaning, Complexity: The Epistemological Premises of Niklas Luhmann's 'Sociological Enlightenment.'" *Philosophy of the Social Sciences* 16 (1986): 115–27.

Contributors

LINDA BRIGHAM is an associate professor of English at Kansas State University. She is currently researching the functionality of the autobiographical and narrative self in a larger framework of alternative functional identities.

BRUCE CLARKE is a professor of literature and science in the department of English at Texas Tech University and a past president of the Society for Literature, Science, and the Arts. His books include *Energy Forms: Allegory and Science in the Era of Classical Thermodynamics* (Michigan, 2001); the collection *From Energy to Information: Representation in Science and Technology, Art, and Literature*, co-edited with Linda Dalrymple Henderson (Stanford, 2002); and *Posthuman Metamorphosis: Narrative and Systems* (Fordham, 2008).

MARK B. N. HANSEN is a professor in the program in literature at Duke University. He is the author of *Embodying Technesis: Technology beyond Writing* (Michigan, 2000); *Bodies in Code: Interfaces with Digital Media* (Routledge, 2004); and *New Philosophy for New Media* (MIT, 2006).

EDGAR LANDGRAF is an associate professor of German at Bowling Green State University. Drawing on contemporary systems theory, he has published articles on the Enlightenment, German Romanticism, Goethe, Kant, Nietzsche, Derrida, and Luhmann. He is working on a book with the working title "Improvisation, Art, and the Art of Living: Unpredictability and Creativity in the Age of Goethe."

IRA LIVINGSTON is a professor and the chair of humanities and media studies at Pratt Institute. He is the author of *Arrow of Chaos: Romanticism and Postmodernity* (Minnesota, 1997) and *Between Science and Literature: An Introduction to Autopoetics* (Illinois, 2006). He has co-edited the collections *Posthuman Bodies*, with Judith Halberstam (Indiana, 1995), and *The Poetry and Cultural Studies Reader*, with Maria Damon (Illinois, 2009).

NIKLAS LUHMANN (1927–98) was until 1993 a professor of sociology at the University of Bielefeld, Germany. He was one of the most prominent thinkers in sociological systems theory and the most successful adaptor of the concept of biological autopoiesis to social systems. A prolific writer, his works include *Social Systems* (Stanford, 1995); *Die Gesellschaft der Gesellschaft* [The Society of Society] (Suhrkamp, 1997); and *Art as a Social System* (Stanford, 2000).

HANS-GEORG MOELLER is a senior lecturer in the philosophy department at University College Cork, Ireland. He is the author of *Luhmann Explained* (Open Court, 2006), *Daoism Explained* (Open Court, 2004), and *The Philosophy of the Daodejing* (Columbia, 2006). His most recent publication is a treatise based on Daoism and Luhmann, *The Moral Fool: A Case for Amorality.*

JOHN PROTEVI is an associate professor of French studies at Louisiana State University, and the founding editor of the series *New Directions in Philosophy and Cognitive Science* with Palgrave Macmillan. He is the author of *Time and Exteriority: Aristotle, Heidegger, Derrida* (Bucknell, 1994); *Political Physics: Deleuze, Derrida and the Body Politic* (Athlone, 2001); and *Political Affect: Connecting the Social and the Somatic* (Minnesota, 2009).

MICHAEL SCHILTZ is a postdoctoral fellow of the Research Foundation–Flanders and is affiliated with the department of Oriental studies of the University of Leuven (Belgium). Schiltz's publications include analyses of the media of money and power. He has a particular interest in epistemological and heuristic applications of Spencer-Brown's calculus of indications.

EVAN THOMPSON is a professor of philosophy at the University of Toronto, working in the areas of cognitive science, phenomenology, and philosophy of mind. He is the author of *Mind in Life: Biology, Phenomenology, and the Sciences of Mind* (Harvard, 2007); and co-author of *The Embodied Mind: Cognitive Science and Human Experience,* with Francisco J. Varela and Eleanor Rosch (MIT, 1991).

FRANCISCO J. VARELA (1946–2001), a Chilean biologist and phenomenologist, made pioneering contributions to biological systems theory, immunology, neuroscience, and cognitive science. His works include *Principles of Biological Autonomy* (North Holland, 1979); *The Tree of Knowledge: The Biological Roots of Human Understanding,* with Humberto Maturana (Shambala, 1998); and *Ethical Know-How: Action, Wisdom and Cognition* (Stanford, 1999).

CARY WOLFE is the Bruce and Elizabeth Dunlevie Professor of English at Rice University. His publications include *Critical Environments: Postmodern Theory and the Pragmatics of the "Outside"* (Minnesota, 1998); *Animal Rites: American Culture, The Discourse of Species, and Posthumanist Theory* (Chicago, 2003); and the edited collection *Zoontologies: The Question of the Animal* (Minnesota, 2003). He is founding editor of the series *Posthumanities* at the University of Minnesota Press.

Index

BRUCE CLARKE is a professor of literature and
science in the department of English at Texas
Tech University.

MARK B. N. HANSEN is a professor in the program
in literature at Duke University.

Library of Congress Cataloging-in-Publication Data
Emergence and embodiment : new essays on second-
order systems theory / edited by Bruce Clarke and
Mark B. N. Hansen.
p. cm. — (Science and cultural theory)
Includes bibliographical references and index.
ISBN 978-0-8223-4581-7 (cloth : alk. paper)
ISBN 978-0-8223-4600-5 (pbk. : alk. paper)
1. Cybernetics.   2. System theory.   3. Autopoiesis.
I. Clarke, Bruce, 1950–   II. Hansen, Mark B. N.
(Mark Boris Nicola), 1965–   III. Series: Science and
cultural theory.
Q310.E44 2009
003'.5—dc22
2009012702